W9-DJB-458

CLIMATE
BALANCE

CLIMATE
BALANCE

A Balanced and Realistic
View of Climate Change

STEVEN E.
SONDERGARD

TATE PUBLISHING & *Enterprises*

Climate Balance
Copyright © 2009 by Steven E. Sondergard. All rights reserved.

No part of this publication may be reproduced, stored in a retrieval system or transmitted in any way by any means, electronic, mechanical, photocopy, recording or otherwise without the prior permission of the author except as provided by USA copyright law.

The opinions expressed by the author are not necessarily those of Tate Publishing, LLC.

Published by Tate Publishing & Enterprises, LLC
127 E. Trade Center Terrace | Mustang, Oklahoma 73064 USA
1.888.361.9473 | www.tatepublishing.com

Tate Publishing is committed to excellence in the publishing industry. The company reflects the philosophy established by the founders, based on Psalm 68:11,
"The Lord gave the word and great was the company of those who published it."

Book design copyright © 2009 by Tate Publishing, LLC. All rights reserved.
Cover design by Blake Brasor
Interior design by Nathan Harmony

Published in the United States of America

ISBN: 978-1-60799-598-2
1. Science: Environmental Science
2. Political Science: Public Policy: Environmental Policy
09.11.05

WHAT ARE THE GUIDING PRINCIPLES AND THE DESIRED APPROACH FOR THIS BOOK?

In researching the subject of climate change, the author was troubled by the significant polarization and lack of total honesty found in the rhetoric on climate change from opposing points of view. The two polar positions usually start with one of two assertions: either that significant global warming is indeed occurring and is due to human causes, or that global warming is not occurring; or if it is occurring, it is simply due to natural causes. One's starting assertion is then often supported by one-sided data. This is all too common. The author found virtually no available literature that took a balanced or realistic view of the subject.

As a balanced perspective appeared unavailable, the author started compiling his own database of facts from both sides and began to weigh the various arguments. Many sources were reviewed and weighed while seeking a balanced view. The database grew throughout 2007 and into 2008, during which time it was shared with others who provided additional sources and feedback. The author did not perform any original research for this work. The sources can be found referenced throughout the book and in the bibliography. The content of this work is dominantly the edited compilation of the work of others, which the author has analyzed, filtered, organized, and upon which additional calculations were performed. Independent conclusions then began to form. This work compiles multiple views, plus points and counterpoints on several aspects of climate change. Additional calculations and crosschecks were performed by the author to either verify claims or to gather perspective. The author essentially wrote the book that he hoped he would have found on the subject. Yet, do not expect it to be fully exhaustive. There is far too much information available for a single work to be exhaustive regarding this topic. This book does cover the major aspects, however, and it is believed to include enough pertinent information to provide the reader with a balanced and realistic view. The guiding principles for the work are:

- First, try to set aside all paradigms and initial beliefs on the subject to become as objective as possible.
- Second, gather data from many sources and from opposing sides of the argument.
- Third, establish the potential cause-and-effect relationships and develop a sequence of questions that would best be answered.
- Next, put the data together as one would a jigsaw puzzle of information.
- Assume that most data has at least a thread of truth, the whole truth is rarely relayed, and statements are often slanted toward a specific conclusion.
- Finally, see where the data leads.
- The analysis and conclusions of this work should not be viewed as carved in stone. There will undoubtedly be new data arising and made available after this work is finished. When this occurs,

the analysis and conclusions could very well change. What is presented here can only be based on the available information at the time of writing—2008.

- It is hoped that this work will stretch the reader's thinking a bit by presenting a balanced view and the information might provide increased insight into the subject.

This book is therefore dedicated to as much honesty and integrity as is realistically possible on the subject.

TABLE OF CONTENTS

INTRODUCTION

Global warming and climate change are related subjects that are believed by many to be among the largest societal issues today, and they are now receiving considerable attention by scientists, the media, the public, and politicians. A huge swell in research is currently being funded, and the available information changes rather frequently. It is very hard to keep abreast of this new information, much less analyze it, filter it, and weigh it to achieve any sort of conclusion. It can be observed that most of the media, politicians, and academia do not take the time for this analysis, and they usually report on the subject from one of two extreme positions. Interestingly, many people have already determined that they are on one side of the argument or the other, having decided this based on a limited understanding of available information. It is only now that people are taking a closer look at opposing viewpoints and the breadth of facts behind climate change. This book is an attempt to organize and unveil multiple specifics, to present disparate viewpoints in order to provide a good balance, to relay the subject at a level where a person of some intelligence can gain an overarching perspective, and to ascertain a well-founded and realistic view. What is requested of the reader is to be objective about what is presented and to have an open mind about changing some of your beliefs, when it might be warranted.

Even though the subject of climate change is complex, there should be a few basic questions that could be asked about climate change and global warming that if answered in an impartial fashion should provide a well-rounded perspective on the subject. Indeed, that is the premise of this book. A set of simple questions are asked and then answered in a sequence that builds upon preceding answers. Each question with its subsequent answer(s) delineates a separate chapter. Asking the right questions is important. As a short summary and for quick reference, multiple aspects of each answer are summarized in bulleted highlights following the question at the beginning of each chapter.

Following these bulleted highlights are more details about some of the aspects in order to gain a deeper understanding of each question's answer. A few of the details include some algebra and some geometry for further clarity, but this math is kept to a minimum, and the math can be skipped without losing the point. An attempt was also made to provide a high-level picture in addition to providing exposure from multiple details. It is believed that this should help provide more balance to the facts and achieve better perception. The attempt at assembling a reasonable balance of views has undoubtedly resulted in the presentation of some information that will be proven later to be incorrect. But, since it is not yet known what information will eventually be shown to be inaccurate, this concern should not be used as an excuse for shunning divergent views without objective evaluation.

This balanced approach is in sharp contrast to other information available on the subject, as it is far too common for the subject of climate change to be approached with a predetermined conclusion from one of the two contrasting positions. People from both sides hold some very strong opinions of the subject matter, and it is frequently charged with high emotion. It is amazing just how often these strong opinions and high emotions are not backed up with an objective assessment of balanced information. People would rather

STEVEN E. SONDERGARD

listen to arguments that support their present paradigms than listen to divergent views that might also contain some truth. This alone is an interesting study in human nature. It is much easier for most people to fervently support something untrue than to break their foundational paradigms. Others approach the subject with ignorance or with an agenda. Because these approaches are common, the truth is not given much of a chance to surface. In fact, it is difficult to find an encompassing reference that even *attempts* to be balanced in its approach. That is unfortunate, and it does not represent much intellectual maturity. The term *intellectual maturity* is used here for the ability to hold two or more of these uncertain and opposing thoughts in one's mind and incorporate each to either improve one's understanding or reach a better conclusion.

Scientists know there is a physical mechanism that acts to increase atmospheric and surface temperatures. This mechanism is called the greenhouse effect, and it can be generally quantified for our globe. Carbon dioxide (CO_2) is known to be a strong greenhouse gas. The concentration of CO_2 and other greenhouse gases are trending upward in the atmosphere. Modern industrial society is emitting large quantities of CO_2 and other greenhouse gases into the atmosphere. The quantity of these man-made CO_2 emissions can be calculated. Increasing atmospheric CO_2 levels can be correlated with increasing global industrialization and fossil fuel combustion. There is evidence that atmospheric CO_2 levels are higher today than at any time in the last few hundred thousand years. In addition, there is strong support that our globe has been warming over the last century, and this has been correlated with an increasing atmospheric CO_2 concentration. Further, Arctic ice sheets and mountain glaciers are observably receding, and ocean levels are calculated to be rising. Our climate does indeed appear to be changing, and it is humans who are being blamed for causing much of this climate change. These carefully worded statements are widely accepted, and they are mostly uncontested.

This book does not contest the preceding statements either. For some, these aspects might be all they feel they need to know. After all, these statements by themselves appear convincing enough, and even conclusive. There is indeed an underlying scientific mechanism; the data does point to a problem, and there is a culprit in CO_2. Others contend, however, that the wording of these statements and the unstated aspects are misleading. There is still information that needs to be discovered about the driving forces of climate. It is widely acknowledged that the greenhouse effect is required for life to exist as we know it on Earth. Water vapor is also known to be a strong greenhouse gas, and its atmospheric concentration is much higher than CO_2 or any other greenhouse gas. Industrial emissions of greenhouse gases have been declining in the United States for over a decade. Factors other than greenhouse gases have been better correlated to increasing global temperatures and climate change. There is evidence that atmospheric CO_2 levels are much lower today than during most of the last few hundred *million* years. The underlying data for global warming is under scrutiny and has been questioned by scientists. Some of the recent global warming is attributed to natural causes. Each of these statements is mostly uncontested as well. In addition, there is evidence indicating that receding Arctic ice sheets and mountain glaciers plus rising ocean levels could be expected from natural causes. Some even convey that climate change due to human causes may not be a major threat.

So what is the real story? Taking action using information from only one perspective or the other might be potentially very dangerous for our society. Information and arguments from opposing sides are therefore examined in relative detail within this book. There are still incongruent aspects concerning global warming that the scientific community needs to wade through. There is a large scientific community that acknowledges our globe is currently warming and that it is not solely the result of natural causes. On the other hand, there are several natural causes that do impact our climate.

STEVEN E. SONDERGARD

The largest uncertainties today revolve around the relative *magnitude* of these natural and anthropogenic (human-caused) impacts and the potential outcomes we humans are likely to cause. The only way to accurately determine the relative magnitudes driving global warming is with a balanced, realistic, objective examination. These examinations and their conclusions are what this book attempts to flush out. Note: the term *Global Warming* (with capitalized letters) is sometimes used in this work to designate dire anthropogenic global warming and to specifically distinguish it from global warming that may not necessarily be anthropogenic or dreadful.

The first several chapters of this book and their respective questions cover scientific material. The science is important, but do not be overwhelmed by the details. These chapters, and answers to their basic questions, are valuable as a foundation of understanding that will be built upon later in the book. The information is laid out in such a manner that a reasonable summary can be gleaned from the Quick Reference sections, with more detail available in the sections that follow. Any given section on more detail can be skipped if either familiarity is high or interest is low. Even if details are skipped in an initial reading, there can be some assurance that they are available for potential reference at some future point in time.

1. WHAT IS CLIMATE, AND WHAT CAUSES IT TO CHANGE?

Focus: Regional climates and average global climate are both defined and described with the major influencing factors that impact the climate.

Layman's Brief:

Climate is simply the average weather within a given region during a period of time, usually averaged over a few decades. The period of time used should be long enough to have some statistical significance, but it also seems appropriate that it should be short enough that the average lifespan of a person living during the period would first recognize the described climate as being representative and second might recognize a change in the climate. Obviously, there are many zones of climate around our globe, and climate varies from region to region. Sometimes it is regional climate that is referenced when climate change is discussed, but more often it is the global climate that is inferred. Global climate is the average of all of the regional climates around the world at any given time.

What has been discovered over the last two to four decades is that our climate has varied to a much greater degree than we had previously realized. Sometimes one or more regional climates have changed in a seesaw pattern with one or more other climate regions, but the average global climate did not change much. At other times, the average global climate has changed, and the regional climates tracked the global climate in a (generally) similar manner. The recent increase in worldwide concern about climate change is both about a change in the average global climate as well as about changes in regional climates.

The influences of our climate are many and varied, and it cannot be described simply or succinctly. It will take the first few chapters of this book to describe the larger of these influencing factors. However, several of the major factors can at least be mentioned at this point: it should not be unexpected that the Sun provides essentially all of the energy that drives much about our Earth's climate. Part of the Sun's energy is reflected back into space, and the amount of this reflection is an influencing factor of our climate. This reflected sunlight is called albedo, and it can occur by the reflection from objects at the surface or from clouds or even particulates in the atmosphere. The energy from the Sun that is not reflected is absorbed by objects at or near the surface. This absorbed energy is what primarily drives our weather and impacts our climate. In addition, any short-term and long-term variation in the Sun's solar output has an impact on our climate as well. The Earth's surface also radiates long-wave radiation that escapes back into space. Some of this long-wave radiation is captured by the presence of certain gases (called greenhouse gases) within our atmosphere, which in turn increases the atmospheric temperature and the surface temperature.

It also turns out that our Earth's various orbital oscillations are quite significant in influencing our climate over long periods of time. Even though these orbital variations do not change the amount of average sunlight that hits the Earth, they do impact when and where

STEVEN E. SONDERGARD

the sunlight hits the Earth. There is substantial and extended evidence of these long-term orbital variations in our past, and we also have the ability to forecast some of their influence on future climate.

There are many secondary effects that further influence our climate. These secondary effects are called *feedback mechanisms*. These feedback mechanisms can have either a positive or negative impact on global warming. Each of these influences, and more, are described in the next few chapters.

2. WHAT ARE "GREENHOUSE GASES"?

Focus: The main irradiative influences on our atmospheric and surface temperatures are shown and discussed. The greenhouse effect is explained, and the significant gases that exhibit this characteristic are listed. A total net warming from the greenhouse effect upon the Earth is then provided, and it is supplied from several sources in order to provide a balanced view.

Layman's Brief:

The term greenhouse gas (GHG) is used quite often and is frequently misunderstood by the general public. Put simply, greenhouse gases are gases that capture heat in the atmosphere. Many chemical compounds present in Earth's atmosphere behave as a "greenhouse gas." These are gases which allow most of the direct sunlight (predominantly visible and shorter wavelengths) to reach the Earth's surface unimpeded. As the direct sunlight heats the surface, infrared (longer wavelengths) energy is re-radiated back to the atmosphere. Greenhouse gases absorb some of this re-radiated energy, thereby allowing less heat to escape back to space, trapping it in the atmosphere. Greenhouse gases present in the atmosphere thus function to reduce the loss of heat into space and contribute to increased global temperatures via this greenhouse effect. More detail will be supplied about this effect shortly.

It should be understood that greenhouse gases are essential to maintaining the temperature of the Earth and without them the planet would be so cold as to be virtually uninhabitable. However, it is suspected by some that an excess of greenhouse gases could also raise the temperature of the planet to uninhabitable levels. The concern surrounding greenhouse gases in the Earth's atmosphere is not the existence of greenhouse gases but the concentration (quantity) of greenhouse gases in the atmosphere.

Quick Reference:

- Greenhouse gases are almost exclusively gases with three or more atoms in their molecules. Molecules with three atoms are termed *tri-atomic*. Every greenhouse gas absorbs electromagnetic radiation at very specific wavelengths in the near infrared spectrum (~0.7 to 30 micrometers). This near-infrared absorption is the property that gives greenhouse gases their designation.
- Examples of greenhouse gases include carbon dioxide, ozone, and water vapor. Carbon dioxide absorbs radiation at a wavelength of 15 micrometers (μm), ozone absorbs radiation at a wavelength of 10 μm, and water vapor absorbs radiation at a wavelength of 7 μm. For reference: the visible spectrum of electromagnetic radiation ranges from just under 0.4 μm for violet light to just over 0.7 μm for red light. Only this relatively narrow spectrum is detectable by human eyes.
- When any tri-atomic or higher-atomic molecules are present as gases or vapors in the atmosphere, they keep some heat from escaping into space and thereby keep the Earth warmer than if they were not present in the atmosphere.
- The amount of warming due to *all* greenhouse gases in Earth's atmosphere is some 31°C to 40°C (about 60°F to

70°F). Therefore, greenhouse gases in the Earth's atmosphere are valuable to life on Earth, because the Earth's surface would be much colder (a frozen tundra) without them.

More Detail:

Greenhouse Gas Absorption of Energy

The Sun produces radiation that can be thought of as primarily distributed across the visible and higher-energy wavelengths of the electromagnetic radiation spectrum. A small portion of this energy reaches the Earth. Some of this portion of energy is reflected back into space by clouds, by atmospheric particles, and by various reflective objects at the surface, such as snow. The amount of energy that is not reflected is absorbed by the surface or some object at or near the surface. These objects and the surface then also re-emit lower-energy electromagnetic radiation as well. The wavelength of this lower-energy electromagnetic radiation is dependent on the temperature of each specific object. This is a well-known property of physics known as *black body radiation*. Because most of the Earth's surface is predominantly within a livable temperature range, the electromagnetic radiation emitted from the surface is principally in the near infrared spectrum. This infrared energy can travel upward from the surface through the atmosphere and then into space. As the individual photons of infrared energy travel through the atmosphere, they could hit a molecule of a greenhouse gas, but the photons can only be absorbed by the greenhouse gas if the photons have a wavelength susceptible to absorption by the specific greenhouse gas it encounters. When this happens, the photon's energy does not escape into space. Instead the photon is absorbed by the specific greenhouse gas molecule and the photon's energy causes the GHG molecule to vibrate, which incrementally heats up the atmosphere. Tri-atomic (and higher-atomic) molecules are the greenhouse gases that absorb infrared electromagnetic radiation (photon by photon) of various wavelengths. The di-atomic gases of nitrogen and oxygen dominate our Earth's atmosphere. Nitrogen does not exhibit a greenhouse gas characteristic, and oxygen exhibits this characteristic only minimally in the infrared spectrum. Most greenhouse gases also absorb electromagnetic radiation in more than one band. Each greenhouse gas usually has unique absorption bands that differ from all other greenhouse gases (methane is an exception), yet absorption bands of different greenhouse gases can overlap.

The physical mechanism of a greenhouse effect, and what prevalent gases cause such an effect, is unequivocal. The biggest questions today revolve around the ultimate impact of the greenhouse effect on climate and how much of that impact is due to man. These questions will be addressed in the course of this work.

STEVEN E. SONDERGARD

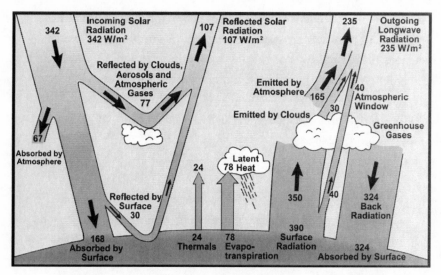

Figure 2.1. The Earth's energy balance on an annual and global mean basis. Each of the numbers represents the makeup of the energy sources impacting this balance, and they are shown in watts per square meter (W/m²). Many of these numbers are not known with high certainty, as they cannot be measured accurately, and some of them are only educated guesses by experts. The left-hand side is basically the incoming solar radiation, and it is balanced by the outgoing long-wave radiation by the Earth on the right-hand side. The incoming solar radiation shown is 342 W/m², with 107 W/m² (31.3%) being reflected back into space. Note that 168 W/m² (about half of incoming solar radiation) is absorbed by the Earth's surface. This absorbed energy is transferred to the atmosphere by thermals (24 W/m²), which warms the air in contact with the surface (by evapo-transpiration, at 78 W/m²), and by long-wave surface radiation that is absorbed by clouds and greenhouse gases (390–324=66 W/m²). (Adapted from: IPCC AR4, 2007, FAQ 1.1 Figure 1, p.94; and Kiehl and Trenberth, 1997)

The Overall Impact of All Greenhouse Gases

The overall temperature impact of all greenhouse gases on the climate of Earth has been calculated by many scientists. There is some variation in the outcomes of these calculations due to specific assumptions, but the cluster of results clearly show that the climate of Earth would be much colder (by these calculated magnitudes) without the presence of atmospheric greenhouse gases. A sampling of these calculated results follows for the total warming from all atmospheric greenhouse gases for our Earth, along with their sources:

40°C—*Rare Earth*, Peter D. Ward and Donald Brownlee, Copernicus, 2000, p.246.

36°C—Calculated by the author using 1372 W/m^2 as the solar constant with a 31% combined albedo, assuming the Earth is in long-term irradiative equilibrium, and then using the Stefan-Boltzmann law, yielding a delta between the average observed surface temperature and the effective emission temperature (290°K–254°K). 36°C is equivalent to 64.8°F.

35°C—Calculated by the author using 1370 W/m^2 as the solar constant with a 30% combined albedo, assuming the Earth is in long-term irradiative equilibrium, and then using the Stefan-Boltzmann law, yielding a delta between the average observed surface temperature and the effective emission temperature (290°K–255°K). 35°C is equivalent to 63°F.

33.3°C—*Thin Ice*, Mark Bowen, Henry Holt and Company, 2005, p.79. (518°R–458°R), or 60°F

33°C—*Physics of Climate*, José P. Peixoto and Abraham H. Oort, Springer, 1992. (288°K–255°K, using 1360 W/m^2 as the solar constant with 30% albedo)

33°C—*The Satanic Gases*, Patrick J. Michaels and Robert C. Balling, 2000, Cato Institute, p.25.

33°C—Richard S. Lindzen, 1992, "Global Warming: The Origin and Nature of the Alleged Scientific Consensus," *Regulation*, p.2. (291°K–258°K)

32°C—NOAA Web site: http://www.ncdc.noaa.gov/oa/climate/global-warming.html#q1, (291°K–259°K)

31°C—Sir Fred Hoyle, 1996. (290°K–259°K)

3. WHAT ARE THE MOST COMMON GREENHOUSE GASES?

Focus: The current and pre-industrial atmospheric concentrations of the most common greenhouse gases are provided. The variability that is observed in some greenhouse gas concentrations is described, and the logarithmic nature of the effect is discussed.

Layman's Brief:

Greenhouse gases are produced by both natural and industrial processes. Several greenhouse gases occur naturally in the atmosphere, such as water vapor, carbon dioxide, methane, and nitrous oxide, while others are only synthetic (man-made). Water vapor is believed to be almost entirely the result of natural causes. Carbon dioxide, methane, and nitrous oxide are both natural and synthetic. The gases that are essentially all synthetic include the chlorofluorocarbons (CFCs), hydrofluorocarbons (HFCs), and perfluorocarbons (PFCs), as well as sulfur hexafluoride (SF_6).

Atmospheric concentrations of both the natural and man-made

greenhouse gases have been rising over the last two centuries due in large part to the industrial revolution.[1,2] As the global population has increased and our reliance on fossil fuels (such as coal, oil, and natural gas) has increased, the emissions of greenhouse gases have risen. The principal greenhouse gases that exist in the atmosphere are:

- **Water Vapor (H_2O):** Relatively high concentrations of water vapor in the atmosphere are the result of the Sun heating the oceans, plants, and soils. Interestingly, water vapors (clouds) are not usually thought of as a bad greenhouse gas or a pollutant.
- **Carbon Dioxide (CO_2):** Natural carbon dioxide enters the atmosphere from the oceanic emissions plus human and animal respiration. Synthetic carbon dioxide enters the atmosphere through the burning of fossil fuels (oil, natural gas, and coal) via waste materials and also as a product of combustion and/or reaction, such as the manufacture of cement. Carbon dioxide is also removed from the atmosphere (or sequestered) when it is absorbed by plants as part of the biological carbon cycle.
- **Methane (CH_4):** Methane is emitted naturally from underground sources but also during the production and transport of natural gas, oil, and even coal. Additional methane emissions result from livestock, rice cultivation, other agricultural practices, and by the decay of organic waste in municipal solid waste landfills. It is also suspected by some that methane emissions might result from living plant matter.
- **Nitrous Oxide (N_2O):** Nitrous oxide is emitted primarily from agricultural activities but can also be emitted by solid waste or even by the combustion of fossil fuels.
- **Fluorinated and Chlorinated Gases:** Hydrofluorocarbons, perfluorocarbons, and sulfur hexafluoride are powerful greenhouse gases emitted primarily from a variety of industrial processes. These gases are typically emitted in much smaller quantities, but they are much more potent.

They are sometimes referred to as high global warming potential gases (high GWP gases).

Quick Reference:

- Water vapor (H_2O) is by far the most abundant greenhouse gas in the Earth's atmosphere (with an estimated average but variable concentration of 1 to 3 percent). Its historic and pre-industrial concentrations are not accurately known.
- Carbon dioxide (CO_2) has an average concentration in the atmosphere at about 0.038%, which is a couple orders of magnitude less than the concentration of water vapor. Note that 0.038% equates to 380 parts per million (ppm). Recent atmospheric analysis by National Aeronautics and Space Administration (NASA) and National Oceanic and Atmospheric Administration, Earth System Research Laboratory (NOAA/ESRL), 2007, indicates that the concentration of CO_2 in the atmosphere varies with altitude. There is also evidence that CO_2 concentrations vary spatially and seasonally. Historically, the atmospheric concentration of CO_2 has cycled between about 180 ppm and 298 ppm as our Earth went through past ice ages. The commonly accepted pre-industrial CO_2 concentration is 280 ppm.
- Methane (CH_4) has a concentration in the atmosphere of about 1.76 parts per million (0.000176%). Pre-industrial concentrations of CH_4 cycled between roughly 0.35 ppm and 0.77 ppm as our Earth went through past ice ages. The U.S. Environmental Protection Agency (USEPA) uses 0.722 ppm for the pre-industrial level. Atmospheric methane concentrations have not risen much since about 1990.
- Nitrous oxide (N_2O) has a concentration in the atmosphere of about 0.32 parts per million. Pre-industrial concentrations of N_2O cycled between approximately 0.20 ppm and 0.28 ppm as our Earth went through past ice ages. The USEPA uses 0.27 ppm for the pre-industrial level.
- Ozone (O_3) has a concentration in the atmosphere in the

parts-per-billion (ppb) range. Ozone is naturally created by lightning and intense radiation in the upper atmosphere. It is also very reactive. Ozone molecules therefore have short atmospheric lifetimes, and its pre-industrial concentration variation is currently unknown.

- Sulfur hexafluoride (SF_6) has a concentration in our atmosphere of about 5.4 parts per trillion (ppt). Its pre-industrial concentration is commonly considered to be 0.0 ppt.
- Carbon tetrafluoride (CF_4) has a concentration in our atmosphere of about 80 ppt. Its pre-industrial concentration was about 40 ppt.
- The absorption of electromagnetic radiation by a GHG is somewhat logarithmic with respect to concentration. This means that further increases in concentration have a reduced impact on warming. This results because there is a finite quantity of photons having the right absorptive wavelengths, and these photons diminish in number as they traverse the pool of atmospheric GHGs.
- Interferences between greenhouse gases (due to overlapping absorption bands) also have an impact on the amount of electromagnetic radiation that is absorbed.
- Additionally, Earth's natural feedback mechanisms affect the final observed warming initiated by the GHGs present in the atmosphere.

Note: there is actually little disagreement with each of these stated concentrations, and they can be found in many sources including: http://www.epa.gov/climatechange/emissions/downloads06/07introduction.pdf, and Spahni, 2005, *Science*, 310: 1317–1321, 25 November (for more detailed historical CH_4 and N_2O values). The pre-industrial concentrations are provided by ice core data, which will be discussed later, in Chapter 5.

More Detail:

The Concentrations and Variability of Common Greenhouse Gases

It is not uncommon for carbon dioxide (CO_2) to be considered the greenhouse gas with the largest impact on climate. CO_2 currently has an average concentration in the atmosphere of about 0.038%. This equates to 380 ppm. Recent atmospheric analysis by NASA and NOAA/ESRL, 2007, indicates the concentration of CO_2 in the atmosphere varies with altitude (from <350 ppm above 30,000 ft to 383 ppm below 5000 ft).[3] Vertical CO_2 variability is corroborated in an article by B. Stephens, et al., 2007.[4] Stephens indicates that summer, mid-day, Northern Hemisphere CO_2 concentrations are generally lower near the surface than in the free troposphere, probably due to greater photosynthesis relative to the anthropogenic emissions. Winter Northern Hemisphere locations were found to have higher CO_2 concentrations near the surface. Locations downwind of continents showed larger gradients than downwind of ocean basins. Southern Hemisphere locations showed relatively small gradients in all seasons. Higher latitudes showed greater CO_2 drawdown at higher altitudes. It is also worth noting that these vertical, seasonal, and spatial CO_2 concentration variations are *not* consistent with the way current atmospheric computer models are constructed. Historically, the atmospheric concentration of CO_2 has cycled between about 180 ppm and 298 ppm over the last few hundred thousand years as our Earth went through the last several ice ages. The commonly-accepted pre-industrial CO_2 concentration is 280 ppm. These concentration numbers are supported by ice core data, and the pre-industrial concentrations can be found in many references, including USEPA and International Panel on Climate Change (IPCC) information.

Water vapor (H_2O), however, has a concentration in the atmosphere of almost two orders of magnitude greater than carbon dioxide. The average concentration of H_2O is around 1% to 3%, but the atmospheric concentration of H_2O is highly variable and even erratic in both the spatial and time domain. This variability should be obvious to any observer of the constantly changing weather in almost any location around the globe. Water vapor, however, is not usually considered in long-term, human-caused climate analyses because it is believed to be 99.9+% the result of natural causes with humans having very little direct impact on its presence in the atmosphere. In addition, its erratic concentration over the surface makes its impacts extremely difficult to determine in long-term atmospheric climate models. Yet, it must be realized that this fact and these difficulties do not reduce the reality that the impact of H_2O as a greenhouse gas is relatively large. An argument is sometimes made that water vapor molecules do not stay in the atmosphere long enough to have a comparable greenhouse effect to CO_2. However, water vapor is continuously and naturally renewed in the atmosphere, and its average concentration in the atmosphere is what determines the magnitude of its greenhouse effect, not the length of time that specific molecules stay in the atmosphere. This is an important fact that is too often conveniently dismissed.

Other greenhouse gases include methane (CH_4) and nitrous oxide (N_2O). Methane (CH_4) has a concentration in the atmosphere of about 1.76 ppm (0.000176%). Pre-industrial concentrations of methane cycled between roughly 0.35 ppm and 0.77 ppm as our Earth went through the past ice ages of the last few hundred thousand years. Nitrous oxide (N_2O) has a concentration in the atmosphere of about 0.32 ppm. Pre-industrial concentrations of nitrous oxide cycled between approximately 0.20 ppm and 0.28 ppm as our Earth went through the past ice ages of the last few hundred thousand years. It is presumed that the concentration of both of these greenhouse gases do not vary significantly with altitude or geography, although this variability does not appear to have yet been analyzed. Even if their concentrations did vary, their relatively low

STEVEN E. SONDERGARD

concentration level in the atmosphere indicates that the impact of any variability in concentration would be small.

Logarithmic Saturation of Infrared Absorption as GHG Concentration Increases

The amount of energy absorbed by a greenhouse gas versus the concentration of the greenhouse gas has a correlation that is logarithmic in nature, indicating the amount of energy absorbed decreases with each incremental increase in the concentration of the greenhouse gas. This is sometimes described as a saturation effect. This effect is described in more detail as follows: as photons having an appropriate infrared wavelength travel through our atmospheric gas mixture, a photon may hit one or more of the greenhouse gases present. When a photon having the right wavelength hits a specific greenhouse gas molecule that can absorb that wavelength, the photon is absorbed, and subsequently the greenhouse gas molecule vibrates with higher intensity, and the atmospheric temperature is thereby incrementally increased. This happens over and over many times with many photons and with many greenhouse gas molecules. As the finite number of photons with absorptive wavelengths traverse the atmosphere, they would be absorbed, their quantity would diminish, and their numbers would approach zero. As the full infrared spectrum of electromagnetic radiation travels through the atmospheric mixture, it therefore becomes more and more depleted of the photons having the absorptive wavelength(s) of the greenhouse gases present.

The logarithmic saturation effect is the simple result of this phenomenon. With incremental increases in the concentration of a specific greenhouse gas, there would be a decreasing number of photons absorbed by the greenhouse gas molecules simply because there would be fewer remaining photons to be absorbed.

It should be noted that the current amount (~380 ppm) of CO_2 in our atmosphere is believed to already absorb most of the absorptive-wavelength photons traveling through the height of our atmosphere from the Sun, plus a significant portion from the surface.[5] This can be observed in Figure 3.1 by examining the direct irradiative greenhouse impact at 380 ppm CO_2 relative to higher concentrations. Therefore, much of the energy available to be absorbed by CO_2 has already been absorbed by its existing atmospheric concentration. Any increases in atmospheric CO_2 concentration would have relatively less direct impact on increasing the atmospheric temperature.

There are additional aspects, such as greenhouse gas interferences and feedback mechanisms that have further impact, and these aspects will be addressed in subsequent chapters.

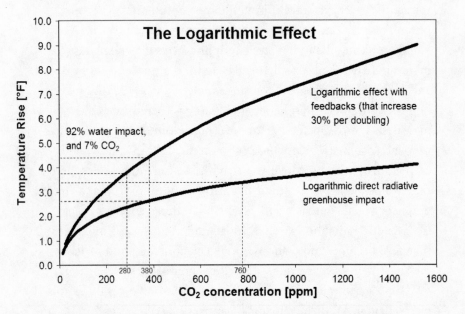

Figure 3.1. This graph is based on a 35°C (63°F) total greenhouse effect, a 7% (4.5°F) CO_2 impact, and a current 1.7 feedback multiplier. Note that doubling the atmospheric CO_2 concentration directly increases the atmospheric temperature by around 1°F, and feedbacks increase that to about 2°F. The logarithmic effect with feedbacks line forwarded here also increases in linearity for each doubling, due to a projected increase in feedback mechanism impacts with concentration.

STEVEN E. SONDERGARD

Interestingly, increasing the pressure of a greenhouse gas mixture would have a similar effect to increasing the concentration, because more of the greenhouse gas molecules would be present within the same volume of gas. (This occurs because it is actually the partial pressure of the gas that directly impacts the absorption, and an increased partial pressure can result from either increasing the concentration or increasing the pressure.) This fact has relevance because we know that atmospheric pressure varies with altitude. Therefore, the amount of energy absorbed by a given greenhouse gas concentration will vary with altitude as well.

4. ARE ALL GREENHOUSE GASES CREATED EQUAL?

Focus: The discussion of absorption bands for the most common greenhouse gases as well as the potency of CO_2 and water vapor is covered. Information on the contribution of water vapor is provided, and from multiple sources, attempting to provide a balanced view. The one-hundred-year global warming potential of the most common greenhouse gases is presented. And lastly, calculations are made of current and pre-industrial CO_2-equivalent concentrations.

Layman's Brief:

The simple answer is that every greenhouse gas is different and they are not created equal. Since each greenhouse gas absorbs only specific wavelengths of energy, each greenhouse gas absorbs different amounts of infrared energy. The atmospheric concentration of each greenhouse gas will also determine the amount of energy absorbed. Higher atmospheric concentrations of a greenhouse gas tend to absorb more of the specific wavelength energy available and thereby reduce the available energy remaining to be absorbed. The atmospheric lifetime of each greenhouse gas can also impact the amount of energy that is absorbed. Chemical reactions can occur in the atmosphere with a greenhouse gas that might produce a reaction product that could further impact the absorption of energy. The atmospheric lifetime, the prevailing chemical reactions, and the overall energy absorbed each impact the global warming potential (GWP) for a relative greenhouse gas concentration. Complex interferences between greenhouse gases also have an impact on the energy absorption.

Quick Reference:

- Water vapor's principle infrared absorption band is relatively wide, from 5.0 to 9.2 μm (with its peak at 6.3 μm). CO_2's principle infrared absorption band is from 14.7 to 16.5 μm. Methane's principle absorption bands are 3.4 μm and 7.6 μm. Nitrous oxide's principle absorption bands are at 4.5 and 7.8 μm. Ozone's principle infrared absorption band is from 9 to 10 μm. This illustrates just some of the differences between greenhouse gases.

- Water vapor and CO_2 are both relatively potent absorbers of electromagnetic radiation in the infrared spectrum. Assuming a uniform spectrum of electromagnetic radiation is available to be absorbed and equal atmospheric concentrations of each greenhouse gas, then CO_2 should absorb between half to twice the energy (molecule to molecule) as water vapor. However, because the actual atmospheric concentration of water vapor is so much greater than other greenhouse gases, water vapor absorbs considerably more energy than any other greenhouse gas, including CO_2, nitrous oxide, and methane. Therefore, water vapor is the most significant greenhouse gas in our atmosphere. CO_2 is considered the most significant *anthropogenic* greenhouse gas.

- Greenhouse gases (such as water vapor and CO_2) can also absorb some electromagnetic radiation in the short wave spectrum (500 to 1,100 micrometers), and there are even some GHGs with absorption bands in the microwave spectrum.

- The United Nations backed International Panel on Climate Change (IPCC) Third Assessment Report (TAR), 2001, estimates the direct and one-hundred-year indirect global warming potential (GWP) for methane and nitrous oxide at 23 and 296 times that of CO_2, respectively (with a stated uncertainty of ±35%). (These multipliers are updated from earlier IPCC work where the GWP for CH_4 was 21 and N_2O was 310, but the factors have not changed since the IPCC TAR.)

- The overall impact of all atmospheric greenhouse gases was relayed earlier as being between 31°C and 40°C. But how much of this total global warming is due to each specific greenhouse gas? The answer varies markedly between literature sources. The following is an example of this disparity. It becomes clear that the Kyoto Report (and other sources) often mislead their audiences by failing to report water vapor in their analysis.

Individual Greenhouse Gas Contributions to Global Warming

	Per Kyoto Report	Per DOE, MIT, and others
CO_2	72%	~2–8%
N_2O	19%	<2%
CH_4	7%	<1%
H_2O	not considered	~90–98%

More Detail:

Here is more detail for those who desire it:[1,2,3] Water vapor's principle infrared absorption bands are relatively wide, from about 5.0 to 9.2 μm (with its peak at 6.3 μm and tapering off by 9.2 μm) and above 14 μm (centered at 65 μm). But water vapor also absorbs radiation in multiple peaks between 0.7 and 4 μm (2.70 μm, 1.87 μm, 1.38 μm, 1.10 μm, 3.20 μm, 0.94 μm, 0.82 μm, and 0.72 μm, in order of decreasing absorption). CO_2's principle infrared absorption band is from 14.7 to 16.5 μm, but CO_2 also absorbs electromagnetic radiation around peaks at 2.7 μm and 4.7 μm. Methane has absorption bands around the peaks at 3.4 μm and 7.58 μm, with smaller peaks at 7.4 μm and 7.87 μm. Nitrous oxide has absorption bands around peaks at 4.49 μm, 7.82 μm, and a smaller peak at 16.98 μm. Carbon monoxide absorbs infrared radiation at 2.3 μm and 4.7 μm. Ozone has an infrared absorption band at 9.0 to 9.6 μm with a smaller peak at 14.2 μm. Ozone also absorbs considerable harmful ultraviolet emissions from the sun below 0.3 μm.

In Figure 4.1, notice that the ground level absorption bands for H_2O and even CO_2 already absorb close to all of the available electromagnetic energy impacting the surface. Compare the observed absorbed energy of H_2O and CO_2 at 11 km (the bottom of the stratosphere) relative to the observed absorbed energy at ground level. The infrared energy emitted from the surface would behave similarly. An increased concentration of either H_2O or CO_2 would therefore absorb significantly less energy than has already been absorbed. It can further be observed that the amount of absorbed energy can vary considerably between greenhouse gases. The absorbed wavelength(s) of a specific GHG correlates with its specific molecular resonant frequencies, which are in turn dependent on the number of molecular bonds, the strength of each bond, and the mass of the atoms being bonded (note that isotopic variation creates multiple peaks). Molecular stretching, twisting, and rocking of the bonds between the atoms can each occur. Higher-atomic molecules obviously have more bonds and potentially more bond types, and natural gaseous mixtures have isotopic variation; therefore, GHG molecules can have many absorption bands, even in the infrared spectrum.

STEVEN E. SONDERGARD

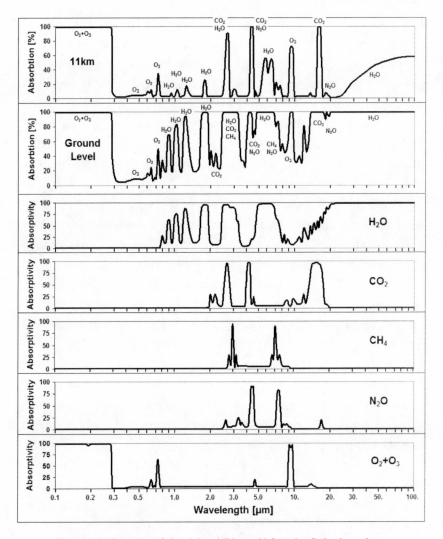

Figure 4.1. Absorption of ultraviolet, visible, and infrared radiation by various gases in the atmosphere. Note the infrared light available at ground level is greatly reduced relative to the bottom of the stratosphere (11 km). The multiple absorption bands for water can be easily observed. The absorption bands can be seen for CO_2 around 2.7μm, 4.7μm, and the principle peak around 15μm. Most of the ultraviolet below 0.3μm is absorbed by ozone and oxygen. (Adapted from: Peixoto and Oort, *Physics of Climate*, 1992, p.93; and http://brneurosci.org/c02.html, Cold Facts on Global Warming)

In addition, the energy contained by a photon is dependent on its inherent wavelength. It can be thought of as the shorter the wavelength the faster the photon is vibrating and the more energy it contains. Across the infrared portion of the electromagnetic spectrum, the energy content of a photon does not vary greatly, but it does vary. A greenhouse gas molecule that absorbs photons having a shorter wavelength will therefore absorb slightly more energy per absorbed photon than a different greenhouse gas molecule that absorbs photons with a longer wavelength.

The combined impact of a mixture of GHGs *cannot* simply be added linearly. Pressure, temperature, compositional interferences, absorption band overlap, and polychromatic radiation all impact the absorption by GHGs in a complex manner. Pressure is dependent on altitude, and temperature is dependent on altitude, latitude, and time of day. (For example: Airplane pilots have long used empirical temperature lapse rates for cooler temperatures at higher altitude. The lapse rate that pilots use for dry air is 5.4°F per 1,000 feet, and increasing humidity reduces the lapse rate to as little as 2°F per 1,000 feet.) Mathematical and statistical models are therefore used in an attempt to segregate the impacts of each GHG in our atmosphere. Because of these complexities, uncertainty still exists around the exact impact of each GHG.

Because of its high and variable concentration, the interference of water vapor with other GHGs poses a specific challenge to determining the impacts of other GHGs. It is widely acknowledged that the interferences between GHGs can reduce the overall greenhouse impact of a specific GHG gas by up to 90% of its projected single-concentration impact. The interference between water vapor and methane is in this range. The interference effect depends greatly on the overlap of the absorption bands and the concentrations of the specific GHGs present. The overlap of the absorption bands for water vapor and CO_2 turns out to be significant as well, and the corresponding interference appears to be close to a 25% reduction of the pure-component CO_2 impact. This can be visually observed in Figure 4.1 by examining the overlap of the absorption bands for H_2O and CO_2. The phenomenon can therefore have a significant effect on the overall greenhouse impact in the atmosphere.

Is water vapor or CO_2 the more potent greenhouse gas? Literature sources can be found claiming either water vapor or CO_2 as being more potent. Examining Figure 4.1 indicates that they should not be too dissimilar. Both molecules are triatomic, with water vapor having two hydrogen-oxygen bonds and carbon dioxide having two carbon-oxygen bonds, and both with similar isotopic variation. Theoretically, they should be comparable in their infrared absorption potency (molecule for molecule), but with each having a unique frequency absorption profile. However, after concentration is considered, there is no argument. Water vapor warms the planet much more than CO_2, and it is the most significant greenhouse gas in our Earth's atmosphere. Water vapor, however, is not primarily the result of human causes. Carbon dioxide is therefore considered the most important *anthropogenic* greenhouse gas, which means CO_2 is probably the biggest knob that man can turn to impact global warming.

The Greenhouse Impact of Water Vapor

The overall impact of all atmospheric greenhouse gases was relayed earlier as being between 31°C and 40°C. But how does this overall global temperature increase break down for each specific greenhouse gas? Water vapor is the most significant greenhouse gas in our Earth's atmosphere. Therefore, quantifying the relative impact of water vapor is most important in determining the impact of all other greenhouse gases. The impact of water vapor on global warming has been calculated by many sources. These calculations mathematically consider the relative concentrations of each greenhouse gas, the number and width of their infrared absorption bands, the absorption band wavelengths, and the overlap of absorption bands with other greenhouse gases, among other factors. The impact of all greenhouse gases other than water vapor must necessarily be distributed within the remainder of the total greenhouse gas impact. The relative GHG impact of water vapor in the atmosphere of Earth, along with the source of the calculation, follows:

98%—Lindzen, Richard S., 1992, 2002, "Global Warming: The Origin and Nature of the Alleged Scientific Consensus," *Regulation*, p.2 http://www-eaps.MIT.edu/faculty/lindzen/154_regulation.pdf http://www.ncpa.org/press/transcript/globalwm/globa12.html

95%—http://www.eia.doe.gov/cneaf/alternate/page/environment/appd_d.html

94%—P. J. Michaels and R. C. Balling, 2000, *The Satanic Gases*, Cato Institute, p.25

93%—Secretary General of the International Association for Physical Science in the Ocean. http://www.dailyutahchronicle.com/news/2004/09/10/opinion/point.counter.point.is.global.warming.hot.air.stevenson-715561.shtml

92%—A lenient determination by the author using 35°C (63°F) total greenhouse effect, 4°F (7%) impact from CO_2, plus 1% impact from other GHGs. If the water vapor impact is less than 92%, it was determined that any log effect line either wouldn't go through the origin or the recent 100 ppm CO_2 increase wouldn't sit on the line.

90%–92%—Norman J. MacDonald, "Carbon dioxide is about 5% to 8%," http://www.ncpa.org/press/transcript/globalwm/globa12.html

60%–70%—inferred from 2007 IPCC Fourth Assessment Report (AR4), Working Group I, Summary for Policymakers, p.12, "a doubling of CO_2 concentrations … a best estimate of about 3°C"

If water vapor and carbon dioxide are assumed to have equal infrared absorption potency (molecule for molecule) and the average atmospheric water vapor concentration is 1% to 3%, the greenhouse gas impact of water vapor would be between 92% and 98%. Reports of less than an 84% overall greenhouse gas impact for the water vapor in our atmosphere appear to be based on the implied effect from the AOGCM-based IPCC climate sensitivity with a reasonable logarithmic effect applied.

What Is One-Hundred-Year GWP?

The global warming potential (GWP) of a greenhouse gas is defined as the ratio of the time-integrated irradiative effect from the instantaneous release of 1 kilogram (Kg) of a substance relative to the instantaneous release of 1 kilogram (Kg) of a reference gas.[4] The reference gas is usually CO_2, and the integration time horizon is usually one hundred years. The atmospheric lifetime of a greenhouse gas is therefore quite significant in determining its GWP, in addition to its absorption bands.

Various gases in our atmosphere can have a direct and/or indirect impact on the greenhouse effect. If the gas in question is a greenhouse gas itself, then it will have a direct effect by directly absorbing infrared radiation itself. The magnitude of the direct effect is dependent on the atmospheric lifetime of a greenhouse gas. However, there are also indirect effects. For instance, a gas could chemically react over time with other gases in the atmosphere and transform into a different vapor (greenhouse gas or not). If the product(s) of any chemical reaction alters the irradiative balance of the atmosphere, then there is an indirect effect. CO_2 is quite stable chemically, but methane is believed to chemically destruct in the stratosphere and produce a small amount of water vapor, which is also a greenhouse gas.

Table 4.2 The 100-yr GWP for various gases

	Chemical Name	GWP *	Current Level	Pre-Industrial
CO_2	Carbon Dioxide	1	380 ppm	280 ppm
CH_4	Methane	23	1.77 ppm	0.72 ppm
N_2O	Nitrous Oxide	296	0.32 ppm	0.27 ppm
HFC-32	Difluoromethane	650		~0 ppt (est)
HFC-125	Pentafluoroethane	2,800		~0 ppt (est)
HFC-143a	1,1,1 Trifluoroethane	3,800		~0 ppt (est)
CF_4	Carbon Tetrafluoride	6,500	80 ppt	40 ppt
C_4F_{10}	Decafluoro Butane	7,000		~0 ppt (est)
C_2F_6	Hexafluoro Ethane	9,200		~0 ppt (est)
HFC-23	Trifluoromethane	11,700	14 ppt	~0 ppt (est)
SF_6	Sulfur Hexafluoride	23,900	5.4 ppt	0.0 ppt

(sources include: IPCC 2001 Third Assessment Report, IPCC 2007 Fourth Assessment Report, USEPA, http://www.epa.gov/climatechange/emissions/downloads06/07introduction.pdf)

* According to the IPCC, these GWP factors have a typical uncertainty of ±35%.

Table 4.3 Attributes of Various Greenhouse Gases in our Atmosphere

GHG	Atmospheric Lifetime	Rate of Change *
Carbon dioxide	50–200 years	1.9 ppm/yr
Methane	10–12 years	0.005 ppm/yr
Nitrous oxide	~114 years	0.0007 ppm/yr
SF_6	~3,200 years	0.23 ppt/yr
CF_4	>50,000 years	1.0 ppt/yr

(source: USEPA, http://www.epa.gov/climatechange/emissions/downloads06/07introduction.pdf)

* calculated over recent eight to fourteen year time spans usually starting in 1990; CO_2 is the ten-year average from 1995 to 2005 per IPCC.

It is worth noting that the pre-industrial levels shown in Table 4.2 are what are used for the atmospheric concentrations just prior to the industrial revolution, representing the years around 1800 or so. However, it should be noted that these pre-industrial levels listed are not the highest concentrations known to exist in the last 700,000+ years and prior to the industrial revolution. The highest atmospheric concentration for CO_2 recorded in our ice core data was about 325,000 years ago (corresponding to Marine Isotope Stage 9.2) at 297 ± 3 ppmv. The highest atmospheric concentration for CH_4 from our ice core record is 773 ± 15 ppbv (at MIS 9.3, about 330,000 years ago). The highest atmospheric concentration for N_2O from our ice core record is 278 ± 7 ppbv (at MIS 15.1, around 545,000 years ago). Most of this data is provided by R. Spahni, 2005.[5]

CO_2-Equivalence

Using the GWP factors shown in Figure 4.2, a CO_2-equivalent (CO_2-eq) calculation can be made for the atmospheric concentration of all combined GHGs. The *current* CO_2-eq level can be estimated by the following (without correcting for greenhouse gas interferences):

380 ppm for CO_2 + 1.77 ppm (23) for CH_4 + 0.32 ppm (296)
for N_2O + 80x10^{-6} ppm (6500) for CF_4 = **516 ppm CO_2-eq**

(The other GHGs add less than 0.2 ppm CO_2-eq combined, due to their low concentrations.) If we consider only CO_2 and CH_4 in the calculation, the answer becomes 420 ppm CO_2-eq (which is more in line with what is often quoted). The *pre-industrial* CO_2-eq level can be estimated by the following (without correcting for greenhouse gas interferences):

280 ppm for CO_2 + 0.72 ppm (23) for CH_4 + 0.27 ppm (296)
for N_2O + 40x10^{-6} ppm (6500) for CF_4 = **377 ppm CO_2-eq**

All other GHGs have a negligible pre-industrial concentration. If we consider only CO_2 and CH_4 in the calculation, the answer becomes 296 ppm CO_2-eq (which has been reported by some).

The Stern Report, 2006, states the current level of GHGs in our atmosphere is around 430 ppm of CO_2-eq, and prior to the industrial revolution, the level was only 280 ppm. Another source, P. J. Michaels and R. C. Balling, state the current level of GHGs are about 450 ppm of CO_2-eq, compared with a pre-industrial level of 280 ppm.[6] The difference in these basic calculations of CO_2-eq values in most cases is the result of interpreting the complex interferences that occur between greenhouse gases. Yet, this is unlikely to be the explanation in every case. It is apparent that some analysts simply add the concentrations of the various gases (as was done in the simplified calculations above). In other cases, some sources apparently fail to consider N_2O at all in their calculations, and ignoring N_2O on purpose does not make sense. It also appears that some reports treat either CH_4 and/or N_2O inconsistently in their calculations. It seems unlikely that this would have occurred other than intentionally.

In an attempt to compensate for greenhouse gas interferences, a simple assumption will be made that the impact of the effective reduction in greenhouse gas concentration is a flat 25%. This is a result of greenhouse gases interfering with each other, and particularly with water vapor. If the interferences with water vapor are excluded, this simple reduction impact would be even greater. It is actually much more complicated than this because each greenhouse gas has a different interference impact with every other greenhouse gas. However, this simplified assumption reflects the strong interference with water vapor, and it is illustrative. The *pre-industrial* CO_2-eq level then becomes 377 ppm * 75% = **282 ppm**, which is very close to what most report. However, applying the same 25% reduction factor to the *current* CO_2-eq level yields a number significantly less than what most report. The calculation is: 516 ppm * 75% = **387 ppm**. Even if the calculation is performed with much more complication using specific interference coefficients and concentrations, gaps with what is often reported are still apparent. Note that the 25% GHG interferences almost offset the impact of CH_4 and N_2O, which results in the CO_2-eq values being relatively close to the corresponding CO_2 concentration.

For multiple reasons, it therefore appears that either miscalculations or intentional exaggerations are frequently made in the calculation of CO_2-eq levels, and these misleading statements often result in magnifying the anthropogenic impact and slanting the perspective slightly towards a more pessimistic and unrealistic view of either the present or the projected CO_2-eq levels.

STEVEN E. SONDERGARD

5. IS THE EARTH GETTING WARMER? IS GLOBAL WARMING REAL?

Focus: Many historical reference points are provided from our climate record so that a wide perspective can be obtained. In this chapter, recent temperature trends are discussed as well as satellite and surface weather station data. Complex computer climate models are covered. Ice core dating methods are presented with their findings along with the alternative record from cave deposits.

Layman's Brief:

Is the Earth getting warmer? The straightforward answer is that this depends entirely on the means of measurement and time span that is being examined and compared. A disproportionate number of the highest average recorded temperatures within the last 150-year time span has occurred since 1990 (reported by Al Gore)—based on surface weather station data.

Raw satellite measurements of mid-troposphere (10–30 km) temperatures since 1979 indicate significant cooling, and this data is therefore contrary to surface weather station data. This data has more recently been corrected for various influences to show a warming trend.

Climate models are the primary source of information used by the IPCC. Many of the latest atmospheric climate models predict an increase in mid-troposphere temperatures. However, these models have deficiencies and use built-in assumptions. Model shortfalls currently include using uniform atmospheric CO_2 concentrations, minimizing solar impacts, modeling feedbacks too aggressively, excluding cosmic ray impacts, modeling clouds and aerosols poorly, and excluding boundary layer mixing.

The ice core record provides us with a reasonably accurate climate history that indicates there have been many changes in our past climate. This record indicates that many significant warming and cooling cycles have occurred on Earth in the past.

The answer to the question of global warming truly depends on what time period is being compared and what data is used. An answer to the global warming question could actually be either yes or no. The trends of surface weather station data of either the last one hundred years or the last forty years are the data referenced most often to conclude that the Earth is warming.

Quick Reference:

- The average global temperature record over the past 150 years is a composite of data that includes surface weather station data. This record over the last century indicates the average global temperature has risen about 0.7°C (1.3°F) over this period of time.
- Satellites use microwave sounding units (MSU) to represent the atmospheric temperature, and this has shown tropospheric temperature cooling. It is common to correct the MSU data for many aspects. The corrections

STEVEN E. SONDERGARD

are now as large as the climate change signal, so how these corrections are handled is critical. Various universities and agencies follow different procedures for these corrections, and these differences have led to different conclusions.

- Climate models are the primary source of information used by the IPCC. However, these models have significant deficiencies. One such shortfall with these climate models is that they currently use a simplified, uniform, atmospheric CO_2 concentration level rather than known CO_2 concentration gradients. This alone might explain much of the past discrepancy with the mid- to upper-troposphere temperature measurements of satellites. They also minimize the solar impact, and therefore under-predict diurnal temperature variations. In addition, these models do not model clouds very well, nor do they predict many observed climate phenomena.

- Ice core drilling gives us good records of historic temperature trends. This is because the warmer the temperature at which snow forms, the higher is the proportion of oxygen-18 in the snow's oxygen, due to the higher molecular weight and lower vapor pressure of water containing oxygen-18 versus oxygen-16. Ice core dating can be crosschecked by at least four dominant dating methods: 1) visible layers, 2) dating volcanic fallout, 3) electrical conductivity, and 4) ice-isotopic ratios. These ice cores provide us reasonable historical resolution for essentially each year of snowfall, similar to tree rings. From this, scientists know that the Earth has been generally heating up for the last 11,500 years, since the end of the last great ice age, known as the Younger Dryas, which averaged about 5°C (9°F) cooler than the average twentieth-century temperature. Scientists have also determined that the coldest surface temperatures of Greenland during the last great ice age were as much as 40°F ±2°F colder than what we observe today.

- The Earth appears to have been through many warming and cooling cycles and has seen similar, and even more severe, temperature excursions many times in the last

700,000+ years of our ice core records. In fact, the record indicates that past *natural* temperature swings were wildly variable and even jumpy (large changes occurring in less than four years and possibly less than one year) and additionally that the last 10,000 years of relatively steady surface temperatures are quite abnormal relative to the rest of the record. Much of our ice core record is from central east Antarctica at Vostok (1977–1996), in which the data goes back 420,000 years. More recently (2004), the EPICA Dome C site in Antarctica has produced ice cores that now provide us data as far back as 740,000 years. The lesser-known 1992 Guliya ice core from western Tibet by Lonnie Thompson yielded a chlorine-36 age of 760,000 years (this was first published in 1997).

- The historical average glacial to interglacial temperature swings at the poles appear to have been about 18°F, per Vostok data, with mid-latitude mountain peak swings at about 18°F; tropical atmospheric swings near sea level of about 9°F, per Lonnie Thompson; and with the tropical ocean temperature swings at only about 2°F, per Wally Broecker.

- The Earth started warming after the last ice age, but the ice core data suggests there was a period of sharply colder temperatures (approximately 10°F colder in central Greenland) about 8,300 years ago that lasted for just over a century. This cooling event was sandwiched between two warmer periods of higher temperatures than the average twentieth-century temperature.

- The post-glacial climate peak, the hottest temperatures in the last 11,500 years, is a period termed the Holocene Maximum (or Climatic Optimum, or a Hypsithermal), and data suggests it was in the period from 7,800–5,200 years ago and was about 4°–5°F warmer than the average twentieth-century temperature.

- It is known that there was a period of warmer temperatures of about 2.7°F higher than the average twentieth-century temperature, from around AD 900 to about AD 1300, (termed the Medieval Warm Period—and corresponds with the Medieval Maximum of solar magnetic output recorded from AD 1100 to AD 1250). This warm period allowed, for example, the Viking conquests of the North Atlantic, including the colonization of Greenland.

- Scientists know from the same ice core data that there was a cooler period (of about 1.8°F less than the average twentieth-century temperature) from around AD 1300 to about AD 1850 (this is termed the Little Ice Age—and corresponds to the Maunder Minimum of sun spot activity). This cold spell is also supported by many written descriptions of the weather during this period.

- There is a noted recent upward temperature trend in the last one hundred years of recorded temperatures. The magnitude of this upward trend from 1906 to 2005 is generally stated as 0.6°C–0.7°C (1.1°F–1.3°F). The IPCC AR4, 2007, states the 1906 to 2005 global temperature increase was 0.7°C to 0.74°C (which is 1.26 to 1.33°F). This conclusion is further supported by the work of Johannes Oerlemans, who calculated the temperature sensitivity of forty-eight glaciers on five continents to conclude the Earth was warming at 1.2°F (0.67°C) per century.

- The rate of temperature increase is also estimated to be rising at about 0.2°C per decade. (Note: at this rate of change, the peak of the Medieval Warm Period could be matched within about seventy years.)

- Even within the last century there were noted fluctuations in average surface temperature. There was a warming period from 1900 to 1940 and a cooling period from 1940 to 1970. A global warming trend occurred starting in about 1970 until 1998. The cooling period of 1940 to 1970 is speculatively attributed to various causes. These speculated causes include changes in solar radiation, volcanic eruptions, above-ground nuclear testing, increased particulates and aerosols in the atmosphere prior to the Clean Air Act, and poor data.

- It is a relatively short period of time, but there currently appears to be a slight cooling trend (or at least a leveling off) in the average global temperature record that started after 1998 (a strong El Niño year), and that continues through the time of this writing.

- Although 2007 is only one year, preliminary data indicates that this year had a significant reduction in average surface temperature versus the previous year and with the Northern Hemisphere winter of 2007–2008 recording some of the coldest temperatures in decades.

- Therefore, when asked this simple question, without a specified time period for comparison, the response of many scientists today is predominantly, "Yes, the Earth can currently be considered to be getting warmer," based either upon the weather station data of the last century or the last forty-year trend. But this conclusion cannot automatically be extended to conclude that this warming trend is abnormal, or that the main cause of the warming is due to CO_2, or that the main cause of CO_2 is man, or that we should do something about it, or even that it is bad. (These questions will be examined in turn.)

- What appears to alarm some climatologists about the noted warming trend of the last few decades is the fear that the current trend could be abnormal in its cause and/ or that the velocity of the temperature trend may be more rapid than has occurred in the past and/or that man's impact could cause a potential climate jump. What often goes unrealized is that the historical ice core record of past climate suggests that our Earth may be overdue for a climate change to occur naturally.

More Detail:

Surface Temperature Data and Satellite Temperature Data

A disproportionate number of the highest average recorded global surface temperatures within the last 150-year time span has occurred since 1990, as reported by Al Gore.[1] The temperature database used for this conclusion is a composite from many sources (ships, buoys, weather balloons, satellites, ice cores, and weather stations). The data from weather stations is currently gathered by a worldwide network, the Global Historical Climatology Network (GHCN). But there may be a problem with this data. Most of the world's meteorological stations are now located in or near ever-growing metropolitan centers—which are subject to the urban island heat effect and known to have elevated temperatures by as much as a few degrees relative to nearby less-densely populated areas. There was a decline in the number of surface weather stations from over 15,000 in 1970 to around 5,000 in 2000, and most of the dropped stations were rural.

Much of the decline in surface weather stations occurred from a step-change reduction in the number of stations from almost 12,000 stations prior to 1990 to about 7,000 stations shortly after 1990, with 4,000 of these 5,000 eliminated weather stations located in rural areas. The reason for the reduction in surface weather stations was basically cost reduction. The former Soviet Union fell in 1990, and Western nations experienced a recession at about the same time.

Obviously, the reduction of two-thirds of the surface weather station data in the last three decades of the twentieth century would be expected to have a statistically significant impact on the results. This period corresponds with a recorded increase in the global average temperature. It can be observed in Figure 5.1 that there is an inverse relationship between the number of surface weather stations and the global average temperature. It is important to have a consistent set of weather stations that gather data from the same locations for the entire time period being compared. Weather stations that no longer exist might best be removed from all earlier years as well. This would still not yield perfect data, as population centers have grown and warmed, but reasonable adjustments could be made to compensate for this. The selective removal of weather stations would result in a reduction of temperature data points—but a more consistent database would result that would be far less questionable. Many people are unaware of this significant database problem. Some who are aware of the problem admit that the urban island effect would indeed skew the data, but they contend that it should not change the conclusion that the globe is warming rather quickly. That may or may not be true, but others contend that it would tell a completely different story and even change the conclusion.

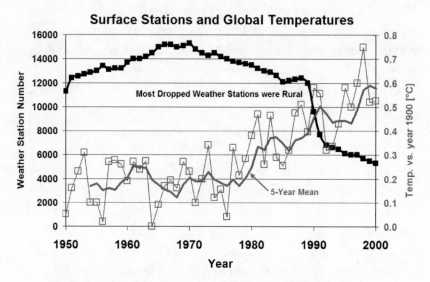

Figure 5.1. Surface station numbers and average recorded temperatures. Note the inverse relationship between global average temperature and surface weather stations. (Sources: The global average temperature is from http://data.giss.nasa.gov/gistemp/graphs/, and the surface weather station numbers are from epw.senate.gov/repwhitepapers/ dcmeetingnov16.pdf, p.8.)

The IPCC says that it is well aware of the statistical problems with the raw temperature data and has taken appropriate statistical measures to correct and clean the data. However, the details of this activity have been shielded from easy scrutiny. Even with cleaned data, the surface weather station number since 1950 versus the global five-year-averaged temperature has a curve fit R^2 of 0.81. Considering this statistical significance, plus the lack of proper statistical measures taken by climatologists with other known climate databases (discussed in Chapter 14), outside scrutiny is probably warranted. Although it would require a large effort, the global average temperature database should be made available, the correction methodologies should be made public, and statistical scrutiny should be performed. This would significantly reduce the uncertainty around a very key aspect concerning global warming.

The first weather satellite went up in late 1978, and satellite measurements since that time suggest there has been significant cooling of the stratosphere (10 km–30 km) temperatures, per the IPCC, 2007.[2] Weather balloon data corroborates the lack of warming in the stratosphere. (Ozone depletion in the upper atmosphere has been forwarded as a reason for stratospheric cooling.) However, the tropospheric temperature trend (up to 10 km) is more uncertain. One group of researchers that interprets and corrects satellite data has suggested there has been a lack of warming in the troposphere since 1979, and there may have been mid-tropospheric cooling relative to the surface. More recent corrections from this group now indicate more of a warming trend. Another group interpreting the same raw satellite data suggests there has been tropospheric warming at all levels up to 10 km. Conclusions have therefore depended on which group, which time, and which corrections are believed to be more accurate.[3] Radiosonde data since 1960 supports warming of the tropical troposphere.[4] Climate models also commonly predict tropospheric warming with maximum warming occurring in the middle to upper tropical troposphere.[5] Most climate models have predicted that as surface temperatures warm the troposphere should also warm and by as much as 1.2 to 1.5 times the surface temperature.

The discrepancies between surface temperature measurements, satellite temperature data, and climate modeling are still being resolved today. There have been several theories as to why the data is inconsistent. One theory is that *surface temperature* measurements are simply misleading us for one or more reasons. The once inviolable global average surface temperature data record is currently under some scrutiny. The drop in surface weather stations and the urban island heat effect in the 1980s and 1990s are only part of the concern. The data within the 1940 to 1970 cooling trend is being questioned, as well as the post-1998 cooling trend.[6] Several portions of the composite database are being individually examined, and corrections are being proposed. Nighttime surface temperatures over land are suspected of showing falsely warmer temperatures. They are considered false because they are not solely the result of an accumulation of heat but are partly the result of turbulent mixing of heat near the surface. This is described as a boundary layer mixing problem, and this effect is not included in present climate models.[7, 8]

Conversely, some have postulated that *satellite data* is where the problem lies.[9, 10] Satellites measure atmospheric microwave emissions by microwave sounding units (MSU). There are now over a dozen satellites that measure these mass-weighted averages. The raw satellite measurements are corrected for non-climatic influences such as inter-satellite biases, calibration errors, changes in instrument body temperature, instrument drift, roll biases, and orbital decays. But the data are also corrected for other aspects. How the data is corrected is critical because the structural uncertainties are now as large as the predicted climate change signal! The raw MSU data has been translated to different temperature datasets by various agencies that follow different data handling methodologies.[11, 12, 13] Carl Mears and Frank Wentz, 2005, highlight one of the alleged errors and their diurnal correction for their Remote Sensing Systems (RSS) dataset.[14] In 2003, the RSS dataset showed a warming of roughly 0.1°C (0.18°F) per decade while the University of Alabama at Huntsville (UAH) dataset showed little change.[15] In 2005, the RSS dataset reflected warming for all three layers of troposphere (consistent with surface temperature measurements), but the UAH dataset showed

STEVEN E. SONDERGARD

cooling in the lower and mid-troposphere.[16] John Christy and Roy Spencer, 2007, who created and maintain the UAH global temperature dataset, now claim that their satellite temperature database shows a warming trend, albeit less than the surface data.[17] Christy and Spencer relay that corrected global temperatures currently show a warming trend of about 0.25°F (0.14°C) per decade since 1979. The IPCC AR4, 2007, conveys that corrected satellite measurements show lower and mid-tropospheric warming rates similar to those of the surface temperature record.

Climate models are another means of determining past temperature trends. The results from these models are compared to surface temperature measurements and satellite temperature data to determine a given model's accuracy. Climate models are continuously being updated in an attempt to make them better in this regard. The IPCC AR4, 2007, claims their models can now predict tropospheric warming up to 10 km and stratospheric cooling from 10 km to 30 km.[18] Climate models will be discussed at more length in the next section.

After examining these many differences, it would not be hard to conclude that discrepancies between surface temperature measurements, satellite temperature measurements, and temperature modeling are still unresolved. One would think that there would not be much discrepancy on these relatively straightforward temperature measurements. Yet, that has not been the case, and a temperature dataset that is uncontested by all concerned is yet to emerge. So how can anyone know which corrections to believe? The disparities point primarily to fundamental deficiencies in the climate models.[19] The divergent opinions indicate that consensus does not yet appear to exist on this seemingly basic aspect, and it leaves scientists with crucial, unanswered questions—at least for now.

General Circulation Models (GCMs)

General circulation models are a class of computer-driven models for weather forecasting, understanding climate, and projecting climate change. They are sometimes called global climate models. These computationally intensive numerical models are based on the integration of a variety of thermodynamics, fluid dynamics, and chemical equations, with some parametric and biological considerations.[20] These models are based on the theory of scientific determinism, which states that the science can be known and modeled to such a degree that the models can be used predictively. The more encompassing atmospheric-ocean GCMs (AOGCMs) are the models that form the basis for the IPCC forecasts. These models are enormous in size and have taken many man years of work to get them to where they are today. They tend to use general rules taken from experience and experimentation, which is to say they use empirical equations—with some parametric and theoretical equations as well. The bigger models require supercomputers to solve, and they have been notoriously unstable in the past. Because these models use general empirical rules, modelers contend that forecasting the future climate (the average weather) is easier than accurately forecasting the local weather even a few days or weeks from now. These models initially took different approaches to one aspect or another, but the widely-used models are now virtually all community efforts that share similar relationships and code. Examples of widely-used community models include the NCAR CCSM3 (National Center for Atmospheric Research, Community Climate System Model 3) and the HadCM3 (Hadley Centre Climate Model 3). It should be realized that AOGCMs are not based solely on science. They use science, but it is important to make the distinction. These models have major built-in assumptions based on invasive paradigms. The results from these GCM models should not be taken as fact. The results are only an indicator of what might occur, completely subject

STEVEN E. SONDERGARD

to the assumptions and correlations of the model itself. These models are simply tools. People who do not understand this can become disconnected from reality.

Clouds, for one, are very hard to model accurately, and averaging the effect of various clouds to achieve an average cloud impact is inherently difficult. Lower altitude clouds act differently than high altitude clouds. To illustrate, low altitude clouds are relatively dense, and they reflect significant solar irradiation during the daytime, while high altitude clouds are less dense and do not reflect much solar irradiation. In addition, the upper atmosphere and polar regions are drier, and a small increase in water vapor in these locations has a greater relative influence on the greenhouse effect. A change in many characteristics of clouds can either amplify or diminish their warming or cooling effect. These changes include altitude, type, location, water content, particle sizes and shapes, and lifetimes.

Better modeling details are limited in GCM models, as larger computers would be required. One heroic group ran their model for a year on a supercomputer. The concept of grid boxes is therefore used to divide the globe into manageably sized chunks. The size and the number of these grid boxes are dependent on the modelers' time and the speed of the computer available. A standard atmospheric grid box size is currently 1.9° latitude by 1.9° longitude, but grid boxes up to 4° latitude by 5° longitude are also used. GCM models cannot really determine what happens inside a specific grid box, and that is why sub-grid parameterization factors are used for these aspects. Parameterization factors are normally used for aspects such as dust, rain, clouds, and even ocean currents. Tornadoes, hail, thunderstorms, lightning, and other similar catastrophic weather occur sub-grid, and they are not modeled well by GCM models. It is quite a stretch to expect GCM models to predict future trends for these catastrophic events.

The vertical dimension wasn't handled well in early GCM models, but current models commonly use nineteen to thirty-one atmospheric levels. The models usually extend to about 30 km above the surface, with the higher levels being thicker. Ocean grid boxes in AOGCM models currently vary between 0.3° to 5° and have from thirteen to forty-seven vertical levels. Stepping through time starting with known characteristics allows the models to project characteristics later in time. A common time-step increment used in modeling is currently thirty minutes.

Modelers have spent entire careers on GCM models. There is therefore great pride and significance at stake in building and sustaining their credibility. Modelers have glowing confidence in their models, and they always report that their models are quite accurate. Of course, accuracy is relative. They cite several reasons for their confidence. 1) The models are based on conservation of mass, energy, and momentum, along with many observed relationships. 2) The models have the ability to simulate past climate variations on various time scales. 3) The models have the ability to simulate important aspects of current climate, such as greater increases in nighttime temperatures, increased polar temperatures, and cooling from volcanic eruptions (e.g. Pinatubo). On the other hand, the diurnal temperature variation (daytime to nighttime temperature swings) is under-predicted by most models on the order of 50%, which could indicate the solar impact might be considerably under-factored within the models and which would translate into factors that are too high for the greenhouse effect. Because GCM models are increasingly a tight-community effort with increasingly interrelated code, the chance that these models might be following an errant group-think mentality is high. The inner workings of the AOGCMs are relatively opaque to all but the most detailed observers. This makes adequate scrutiny of the inner mechanisms of the models rather difficult.

There were twenty-three AOGCM modelings used in the 2007 IPCC AR4 effort, and they are each listed in chapter eight of the IPCC report.[21] In utilizing these models, it is common for modelers to make a series of runs to provide an answer, commonly using about fifty runs in these series. Each run is adjusted for one or more starting conditions to determine sensitivity. This approach allows a probability distribution of answers, in which a confidence level and a statistical range can be determined. The results of the current IPCC/AOGCM models infer an increase of 3.2°C (with a range of 2.1°C to 4.4°C) for a doubling of CO_2 concentration in the atmosphere. This equates to 5.7°F (with a range of 3.7°F to 7.9°F).

The author has extensive computer programming and modeling experience on both the thermodynamic properties of fluids and the optimization models of operations research. This experience has led to an assessment that most anything can indeed be modeled, but that does not mean the model is necessarily an accurate tool or that it is useful for forecasting. That requires the model to be further based on proper fundamental drivers, and these drivers must be modeled with accurate relative strengths and limits. It is quite easy to overcompensate one fundamental mechanism with the under-compensation of another mechanism in a model. But, when that is the case, the model would *not* necessarily be valid or useful in assessing the relative impact of one of these mechanisms. In addition, if multiple models generally follow the same techniques and they are less than accurate for one reason or another, an increased number of models would not add anything to increased credibility or accuracy. These assessments appear to be applicable to the relative state of AOGCM models today. AOGCM models have been able to replicate past temperature profiles with *some* accuracy, but they cannot reproduce many regional aspects, diurnal variances, cloud effects, sub-grid weather, and many observed climate phenomena. The argument has been made that there is too much confidence being placed in these models and that they may be misleading us with incorrect forecasts. A short summary of how well these AOGCMs have predicted observed climate patterns is shown below.

Table 5.2 IPCC AOGCM Predictability

(Source: IPCC, 2007, Fourth Assessment Report, Working Group 1 report, pp.620–629)

OBSERVED CLIMATE PHENOMENON		SUMMARY OF MODEL PREDICTABILITY
Northern & Southern Annular Modes	NAM	"similar patterns", therefore "realistically simulated"
Pacific-North American Pattern	PNA	"replicates various aspects"
Cold Ocean-Warm Land Pattern	COWL	"has been simulated"
El Niño Southern Oscillation	ENSO	some response with "serious systematic errors"
Pacific Decadal Oscillation	PDO	"no difficulty", believed to be part of ENSO
Madden-Jullian Oscillation	MJO	"unsatisfactory"
Quasi-Biennial Oscillation	QBO	models "do not generally include the QBO"
Atlantic Multi-Decadal Variability		"do not exhibit predictability"
Atmospheric Regimes and Blocking		locations are okay, but not frequency or duration
Shorter-Term Predictions		getting better
Extremes of Temperature		"generally well simulated"
Extremes of Precipitation		"less well simulated"
Monsoon variability		"large errors", and appears related to ENSO
Tropical Cyclones		"not resolved"
Summary		"not enough resolution" plus "coarse resolution and large-scale systematic errors" limit modeling

Ice Core Dating Methods and Ice Core Analysis

Many different climate dating techniques are available to scientists. None of these techniques offer as good a glimpse into our past climate as does ice core dating.[22, 23] Written records of volcanic eruptions and other events go back roughly 2,000 years. Living tree rings from bristlecone pines allow us to understand weather patterns back almost 5,000 years. Our complete tree ring record, from logs found in many archeological sites, can be overlapped and matched to now provide us temperature and moisture data back almost 12,000 years. Yet, ice cores from Greenland and Antarctica now provide us a window into our climate and weather patterns over the last 700,000+ years.

After snow falls, it compacts into detectable ice layers that depict the annual variations, very similar to tree rings. These layers have been compared against written events, tree ring records, lake sediments, ocean sediments, and even with other ice cores. The comparison and repeatability of the records provide scientists with great confidence in the accuracy of this record.

STEVEN E. SONDERGARD

The ice layers can be detected by visual means. Warmer, lower density summer snow (called *hoarfrost*) is detectably different than colder winter snow, even after it is compacted into *firn* (German for "old snow"). If the ice core is placed upon a light table, the annual layers are quite visible. The summer layers will appear lighter and the winter layers will appear darker. As the deeper snow layers are compressed by the weight of the snow above them, the ice grains are squeezed together but not squeezed so much that trapped air cannot still move between the ice grains. After a depth of about two hundred feet, equivalent to maybe eighty to two hundred years of snowfall, the *firn* is squeezed sufficiently to trap the air within the ice crystals (this is called *bubble-close-off*), yet summer and winter layers remain distinct. At about 2,000 feet of depth, the ice cores have been squeezed with so much overhead weight that when the ice is brought to the surface it snaps, crackles, and pops as the trapped pressurized air escapes. At about 8,000 feet of depth, the ice crystals actually change from their hexagonal crystal structure to a cubic crystal structure (known as ice 7), and the air molecules slide into the center of the cubic structure. This cubic gas-ice structure is called a gas-hydrate or clathrate-hydrate, and the ice cores are as clear as glass—at least while they are under pressure. After a few days the ice cores become full of bubbles as the pressure is released, and the gas starts to come out of the ice crystal matrix.

But ice core layers can be read by many other means as well. The visible ice layers are first confirmed by crosschecking them against written volcanic fallout records and our tree ring record. In addition, electrical conductivity and ice-isotopic ratios provide further clarification. Because air contains some carbon dioxide, rain or snow falling through the air containing carbon dioxide produces a weak acid called *carbonic acid*. The degree of acidy in the ice changes from summer to winter, and this can be detected when electricity is passed through the ice layers. The layers having more acidity will have increased electrical conductivity. Moving the electrodes down opposite sides of an ice core will yield a continuous measurement of the conductivity of the ice layers. In addition, volcanic eruptions produce a spike in sulfur dioxide within the air, and any rain or snow that falls through this spiked air will include small amounts of sulfuric acid. The volcanic eruptions of Pinatubo in the Philippines (1991), Mount St. Helens (1980 and 1479), Mont Pelée in Martinique (1902), Krakatoa (1883), Mount Tambora (1815), Laki in Iceland (1783), and Vesuvius (79), among others, are easily observed in the conductivity record of the ice cores.

Very small particles in the air, such as aerosols or dust, can also be trapped in the ice layer record. These dust particles trapped in the ice are another means to crosscheck the known volcanic eruptions. The dusts found in the ice cores can also be compared against the many geological formations all over the world, and from this analysis scientists gain insight into the historical wind patterns. The elements of lead, strontium, and neodymium have each been used as tracers to help determine the origination of the dust particles. It should not be too surprising to learn that many of these wind patterns changed when the global temperature changed. Interestingly, sometimes winds from halfway around the world reach the ice sheets. Dust blows in regularly from southern South America to fall on Antarctica. But it was found that dust from Venezuela and even China has periodically blown in to fall on Greenland.

Cosmic rays are single, charged protons that have extremely high energy levels. In fact, the energy content is so large they can split an atom of beryllium. The beryllium-10 and other isotopes, formed in the atmosphere from cosmic rays, fall to the surface, and some of them land on the ice sheets to be preserved in the ice layers. We therefore have a reasonable historical record of the cosmic rays hitting Earth over time. The eleven-year sun spot activity cycle is recorded very nicely in the ice core record. This eleven-year variation in our Sun's output is about 0.1%. Longer-term variability in solar output from our Sun also exists. An eighty-year solar cycle may exist and is at the edge of detection. The Maunder Minimum of sun spot activity plus the Dalton Minimum are both recorded in the ice core layers over more than a century. Most estimates of the Maunder Minimum indicate that a reduction in our Sun's output was about -0.2% during this period, with some estimates at -0.4%, and one estimate at -0.7°C.[24] The Dalton Minimum has been estimated at as much as -0.6°C.[25]

The atmospheric gases trapped within the ice can be analyzed to determine their constituents. The concentration of carbon dioxide, methane, and other gases can be measured at each ice layer to determine their concentrations and how the atmospheric concentrations of these gases have changed throughout the historical record.

Ice-isotopic ratios are very valuable in reading the ice core layers. First a background: Almost every element has naturally occurring isotopes in which the number of neutrons varies within the atom's nucleus. In the case of hydrogen, the number of protons is always one, but the number of neutrons can be zero, one (deuterium), or two (tritium). The atomic weight of these hydrogen isotopes would therefore be one, two, or three respectively. In the case of oxygen, the number of protons is always eight, but the number of neutrons can be eight, nine, or ten. The atomic weight of these oxygen isotopes would therefore be sixteen, seventeen, or eighteen respectively. When these atoms combine to form water, they have the potential to form molecules with an atomic weight of anywhere from eighteen to twenty-two—tritium is unstable and has a half life of only a few years, so water-23 and water-24 would be short-lived. Second, 99.9% of hydrogen is hydrogen-1 and 99.8% of oxygen is oxygen-16; therefore, water-18 is by far the most common form of water. But the other molecular weights of water do exist naturally. Third, the physical behavior of water varies with its molecular weight. Vapor pressure is one of those properties. Even though the quantity is relatively small, the heavier water molecules are more likely to condense from the vapor phase to form rain or snow at a higher temperature. Therefore, the isotopic ratio of the rain or snow will vary with the atmospheric temperature at which it forms.

Isotopic ratios form the basis for determining past temperatures. This science is known as *paleothermometry*.

By measuring the ice-isotopic ratios in each ice core layer, scientists can determine the atmospheric temperature at the time the snow formed. (The ^{18}O content of the ice decreases about 0.04% per 1°F of cooling.) This was first published by Willi Dansgaard in 1964. Measurement errors in this process are thought to be small. The correlation can be affected by humidity, the seasonality of precipitation (influencing cloud heights, humidity levels, and temperature lapse rates), heavy isotopic depletion after heavier snowfalls, and even ice friction within the accumulated snow and ice itself. But the atmospheric temperature has by far the largest impact on this correlation—even in the high-altitude tropical glaciers such as the Andes, where 90% of the snow falls in the wet season. This was made more evident by the Sajama ice cores (published in 1998) compared with the Huascarán ice cores (published in 1995). The Sajama ice was eight times less dusty during glacial times yet had thicker and colder accumulation layers, versus the Huascarán ice cores being two hundred times dustier during glacial times. This indicates that it was wetter at Sajama and dryer at Huascarán during glacial times. But even with this variation, the isotopic ratio correlates well between Sajama and Huascarán and Greenland and Antarctica. Then in 2003, our understanding became even clearer when Katherine Visser of the USGS separated the impact of temperature and ice volume using a seabed core from Indonesia's Makassar Strait, based on the ratio of magnesium to calcium in shells.

There have been many ice core drilling projects that have given us ice core data. The Greenland Ice Core Project (GRIP) was primarily a European endeavor between 1989 and 1992. Seventeen miles away, the Greenland Ice Sheet Project 2 (GISP2) was primarily a U.S. effort between 1989 and 1993. Dye 3 is sometimes referred to as GISP1. Much of the ice core record is from central East Antarctica at Vostok, where several ice cores were drilled between

1977 and 1996. The Vostok ice cores give us data that goes back about 420,000 years. More recently (2004), the European Project for Ice Coring in Antarctica (EPICA) Dome Concordia (Dome C) site in Antarctica has produced ice cores that now provide us data as far back as 740,000 years. Recent radar analysis suggests that other sites, such as Antarctica's Dome Argus (Dome A) may have enough ice depth to extend our ice core record to 1,000,000 years or maybe even 1,500,000 years. This Dome A area is projected to be drilled by a team from China and/or the U.S. within the next two to five years. In addition, there is significant research currently being performed on the biological information within ice cores.

Many ice cores have now been drilled, read, and crosschecked against each other, and with very good agreement. Agreement does not necessarily prove accuracy. But the ice core record of the last 12,000 years compares very well with tree ring data, and this agreement plus the long-term agreement between various ice cores does provide scientists with high confidence that they have a pretty accurate record over the last 400,000+ years, even extending to 700,000+ years. The National Ice Core Laboratory in Denver, Colorado, is the repository for most U.S. ice cores.[26] The repository for the European ice core efforts is at Bern, Switzerland. Lonnie Thompson, from Ohio State University, has a repository for his mid-latitude, high-mountain ice cores.

What Does Our Ice Core Record Tell Us?

It has been discovered that there have been many ice ages in our past. These ice ages have been evenly spaced at about 100,000-year intervals for close to a million years, and they are reasonably similar in magnitude and duration. These repeated cycles have a roughly 90,000-year gradual, although wildly variable, cooling period and then an approximately 10,000-year warm period.

There are two interglacial warm periods in the record that are longer than average duration. One was about 410,000 years ago, at MIS 11.3. It lasted about 25,000 years, and it had a mean temperature comparable to today. The other extended interglacial warm period was around 575,000 years ago that started at MIS 15.1 and was about 27,000 years in length. These warm periods correspond with a time when the orbit of our Earth was closer to being circular rather than more elliptical. (Orbital eccentricity will be discussed in the next chapter.) These two interglacial periods are *much* longer than any other interglacial warm period in our current ice core record. The average interglacial period is only about 10,000 years in duration. The oldest four interglacial periods in our current record were generally cooler but lasted longer than the following three younger interglacial periods. Obviously, we are still in the youngest interglacial period.

In most locations at the poles and in the tropics, the air seems to have turned dustier during glacial times, with the 1993 Huascarán ice cores indicating about two hundred times more dust during glacial times than today. During glacial periods, mid-latitude nitrates dropped by a factor of two or more, pollen counts dropped, dryer air led to lower temperatures (by 14°F to 22°F at 20,000 feet per the Huascarán ice cores—inferring 9°F to 11°F at sea level), and the rain forests seem to have shrunk considerably. A 9°F tropical cooling in glacial times was supported in 1993 by Rick Fairbanks's work on Barbados corals. Then in 1994, Martin Stute and Wally Broecker indicated 10°F of tropical cooling in glacial times by groundwater analysis in lowland Brazil.

Most of the high-mountain mid-latitude glaciers do not contain ice that extends earlier than the Last Glacial Maximum (LGM), about 20,000 years ago. The Sajama (at 21,460 feet and 18° south) ice record from Bolivia extends only 25,000 years. The Huascarán (at 22,130 feet and 9° south) glacial record from Peru extends only 19,000 years. Garganta Col (at 19,800 feet) on Huascarán was also bare at the LGM. The Kilimanjaro (at 19,340 feet and 3° south) ice record extends only 11,700 years. The Dasuopu glacier record on the flanks of Xixabangma (at 26,290 feet and 28° north) from the Himalayas extends only 8,000 years. Why are all these glaciers so young, when temperatures were assuredly lower before their birth than they are today? Lonnie Thompson's asynchrony theory explains that the cyclical precession of the Earth's axis heated the mid-latitude oceans and increased the atmospheric moisture enough to build high-altitude tropical snow. Increased or prolonged global warming would therefore require higher and higher altitudes to retain ice in solid form. A reduction in humidity would also lessen snowfall and increase sublimation, causing glaciers to diminish—as appears to be occurring today.

This theory leads to the conclusion that mid-latitude mountain glaciers are rather uncommon in Earth history, and they only occur during a relatively brief time span starting in the warming period following a glacial maximum and lasting only a few thousand years before vanishing prior to the end of the interglacial periods. Mid-latitude glaciers form only when both the moisture level is high enough and the temperature is low enough, and this situation has seemingly existed only 10% to 20% of the time during the last few million years of glacial and interglacial cycles. Mid-latitude mountain glaciers can therefore be expected to recede and diminish naturally during any late interglacial period. This is an important conclusion. For similar but opposite reasons, the Antarctic ice sheet is currently increasing in size, even with the widely reported calving that has recently been taking place. It would be inappropriate to assert or conclude that either glaciers or ice sheets should be constant in their size or that it is abnormal to observe receding glaciers and ice sheets in the present late interglacial period.

The average polar temperatures during the last interglacial period (about 125,000 years ago, MIS 5.5) were 5.5°F to 9°F higher than the present average temperature. Correspondingly, the sea level during this last interglacial period was likely thirteen to twenty feet higher than the ocean level during the twentieth century, per the 2007 IPCC (AR4) report, or between sixteen to twenty feet higher than today, per Mark Bowen, 2005.[27] This is substantiated by dating the coral deposits that are found above the current ocean level. Obviously, coral only grows below the ocean surface, so this is a good indicator that ocean levels have been higher in the past.

The past climate of Earth has historically been wildly variable, with larger and faster temperature surges than anything we have observed in the last 11,000 years (since the last ice age). In fact, during the last ice age (prior to about 11,000 years ago and extending back about 100,000 years, MIS 2–4), there were over twenty abrupt and dramatic temperature swings that are prominently displayed in the ice core record of Greenland. At least one of these average global temperature swings was as much as 18°F in not more than a few decades and possibly in only a few years.

The large temperature shifts indicated in the ice core records of both Greenland and Antarctica are usually not synchronous and are often observed in the opposite direction. This indicates that a redistribution of the available global heat has occurred frequently in the past. This fact provides us with some clues that will be addressed shortly.

We can observe in the ice core record that about 11,500 years ago our weather suddenly and dramatically changed. It was at this time that our Earth came out of the last great cold surge (termed the *Younger Dryas*) and into the modern warmer climate (termed the *Holocene* and also known as *MIS 1*). In addition, the Younger Dryas started about five hundred years earlier in the Southern Hemisphere, and it was much less intense in South America.[28]

The Earth started warming after the last ice age, but the ice core data suggests there were still changes in the climate. For example, there was a period of sharply colder temperatures (about 10°F colder in central Greenland) starting about 8,400 years ago that lasted for 180 to 200 years. This cooling event was sandwiched between two warmer periods of higher temperatures than the average twentieth-century temperature.

The period from about 11,000 years ago to about 4,000 years ago is known as the African Humid Period. An increase in rainfall appears to have turned the Middle Eastern and African deserts into a veritable Garden of Eden at about 7500 BC to about 4500 BC. Around 6400 BC most of the Earth had a climate shock of cold dry weather that lasted almost two hundred years. The Kilimanjaro ice record shows the windswept ocean salts as a spike in fluorine ions, and the GRIP ice record shows a methane level dip for two hundred years, among other evidence. This is supported by archeological evidence in Levant, Mesopotamia, the Persian Gulf, and southeast Iraq along the Tigris and Euphrates rivers. At about 4000 BC this region became dry enough that the Sahara desert began to appear. About 3000 BC there was another drought that lasted for about one hundred years. This drought corresponds with the collapse of the Uruk civilization. Then again about 2200 BC there was another drought, corresponding with the collapse of both the Akkadian empire and the Egyptian Old Kingdom—which had existed since about 3100 BC. This drought was extensive enough that it is recorded as an inch-thick layer of black dirt at a depth of 106 feet in the Kilimanjaro ice record. This drought also corresponds with a sharp reduction in tree ring widths in North America between 2278 BC and 2056 BC. Peru and the Incan Empire were not immune from the impacts of the 2200 BC climate shock either.

The post-glacial climate peak, the hottest temperatures in the last 11,500 years, is a period known as the Holocene Maximum, sometimes termed the Climatic Optimum, or the Hypsithermal, and data suggests it was in the period from about 7,800–5,200 years ago and was about 4°F to 5°F warmer than the average twentieth-century temperature.

It is known that there was a period of warmer temperatures (of about 2.7°F higher than the average twentieth-century temperature) from around AD 900 to about AD 1300—called the Medieval Warm Period—and this corresponds with the Medieval Maximum of peak solar magnetic output from about AD 1100 to AD 1250. This warm period is attributed to higher solar activity by correlating increased carbon-14, resulting from increased cosmic rays, in the tree rings from this period. The correlation between variable solar output and carbon-14 was first promoted by Jack Eddy from NCAR and later confirmed by Stoiver and others. The Medieval Warm Period allowed, for example, the Viking conquests of the North Atlantic, including the colonization of Greenland.

It is known that there was a cooler period, of about 1.8°F less than the average twentieth-century temperature, from about 1520 to 1880—called the Little Ice Age—which encompasses the Maunder Minimum of sun spot activity (1645 to 1715) and the Dalton Minimum (1795 to 1825). The isotopic ratios of the ice cores such as Huascarán and Quelccaya are among the best for this determination. This cold spell is also supported by many written descriptions of the weather during this period, including the annual freezing of the northern European canals when Hans Brinker skated around on them.

Cave Deposits (Speleothems)

There are several caves around the world that have laid down mineral deposits from ground water that provide a means of studying the past climate over the last ~500,000 years using uranium-series dating and oxygen isotope analysis.[29, 30, 31, 32] These caves include the Devil's Hole in Nevada near the California border, the Hulu cave near Nanjing China, and caves on Socotra Island off mainland Yemen. Cave deposit records generally follow the ice age trends of ice core records, but there are some inconsistencies, and the two camps of investigators have not yet been able to fully resolve the discrepancies. This is mentioned here for awareness that there are alternate analysis mechanisms and even some alternate, albeit mostly minor, conclusions.

The Twentieth-Century Temperature Trends, Including the Reduction from 1940 to 1970

There was a warming period from 1900 to 1940 and a cooling period from 1940 to 1970. A global warming trend started again in about 1970 that lasted until at least 1998. The cooling period from 1940 to 1970 is speculatively attributed to various causes. These various proposed causes include changes in solar irradiation; volcanic eruptions; increased particulates in the atmosphere prior to the Clean Air Act; or above-ground nuclear testing, done from 1953–1962 by the U.S. and the Soviet Union, and 1968–1975 by France and China. Others contend that this mid-century cooling was only a Northern Hemisphere phenomenon (refer to Figure 5.4). The twentieth-century temperature trend actually has a fairly high correlation to the variation in solar irradiance during this time period. This will be shown in the next chapter. Therefore, the most probable cause of this mid-twentieth-century cooling trend appears to be due to the total effect of the Sun.

In addition, one cannot help but conclude that any increase in surface temperatures since 1990 has assuredly been exacerbated, to at least some degree, by reduced particulates in the atmosphere resulting from lower sulfur diesel and gasoline fuels in both the U.S. and Europe. These lower sulfur regulations were enacted around 1990 on both continents. The lower sulfur levels of these fuels have reduced the sulfate emissions, and scientists know that atmospheric sulfates are aerosols that help to cool the atmospheric temperature. The enactment of these regulations is not being judged here. The directional impact is simply being noted, with the realization that these earlier, well-intentioned environmental actions have contributed to global warming since about 1990. This fact might help explain some of the observed surge in global average temperature increase from 1990 to 1998. It also highlights the ever-present existence of unintended consequences to every action measure.

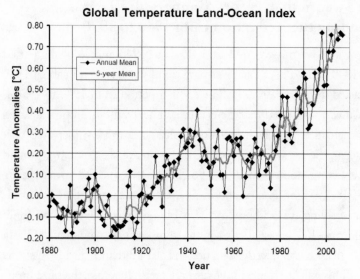

Figure 5.3. The global average temperature from both land and ocean. Note the temperature drop from about 1940 to 1970. (Adapted from: http://data.giss.nasa.gov/gistemp/graphs/)

With the available evidence, most scientists today agree that the Earth can be considered to be getting warmer over the last century and over the last forty years. The majority of scientists also admit that the rate of temperature increase has been relatively rapid. However, it is also admitted that the ice core temperature record indicates that the global average temperature has been warmer in the past and that many rapid temperature excursions have occurred in the past—both from natural causes. The conclusion about global warming cannot automatically be extended to conclude that the present warming trend is abnormal, or that the main cause of the warming is due to CO_2, or that the main cause of CO_2 is man, or that we should do something about it, or even that it is bad. These questions will be examined in the following chapters. On the other hand, a conclusion cannot be made that the current upward temperature trend is solely or even primarily the result of natural causes.

STEVEN E. SONDERGARD

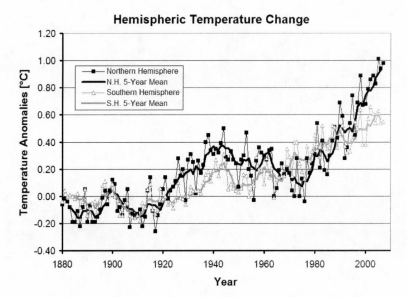

Figure 5.4. The increase in temperature in both the Northern Hemisphere and the Southern Hemisphere. Note that the Southern Hemisphere does show some temperature drop from about 1940 to 1970. (Adapted from: http://data.giss.nasa.gov/gistemp/graphs/)

6. WHAT FACTORS IMPACT CLIMATE, AND HOW MUCH GLOBAL WARMING IS DUE TO THE GREENHOUSE GAS CO_2?

Focus: The Vostok ice core correlation between temperature and CO_2 is presented. The total solar irradiance is discussed, as well as its variability. The orbital periodicities (the Milankovitch cycles) are covered along with various feedback mechanisms and a rough overall feedback effect calculation. Ocean–atmospheric CO_2 balances are mentioned, and ocean currents are described. Dansgaard–Oeschger oscillations, Heinrich events, and Bond cycles are discussed. Finally, a few simple calculations and conclusions are made about the impact of CO_2.

Layman's Brief:

The impact humans may be having on the climate focuses on the last two hundred years since the industrial revolution, with the bulk of the CO_2 and other GHG increases occurring in the last one hundred years. This increase in GHGs is often correlated to the temperature rise of the twentieth century, with the alleged conclusion that the GHG concentration is the cause for the effect of the observed temperature rise.

There are many factors that affect the average global temperature. Among these factors are solar irradiance, albedo, greenhouse gases, aerosols, clouds, solar system orbital variations, ocean currents, volcanic activity, and feedback mechanisms. There is no argument about the existence of these multiple factors. The main debate is about the magnitude of the impact for each factor, the variation in magnitude for each factor, and during which time period(s) each factor has had increased influence.

There is a strong correlation in the ice core record between atmospheric CO_2 concentration and temperature. The Vostok ice core record is commonly cited to show this correlation. Yet, the cause-and-effect relationship is not simple, and the purported conclusion that CO_2 is the cause is not supported by the data. In addition, the quantity of carbon in the atmosphere is only a small amount of the overall carbon that can potentially impact the atmospheric CO_2 balance. Therefore, any changes to the other carbon sources could have a significant impact on the atmospheric CO_2 concentration. The ocean-atmospheric carbon balance is believed to be particularly important.

The orbital variations of the Earth, while not impacting the average solar gain of the Earth, are known to have an immense impact on the global climate. Solar gain is simply a term for the incoming solar energy. The solar output of the Sun (and possibly the variation in the type of its output) appears to be among the most influential of the factors impacting climate. Surprisingly, the solar irradi-

STEVEN E. SONDERGARD

ance reaching the polar latitudes (particularly the northern latitudes around Greenland) seems to be more important to our climate than the solar irradiance elsewhere. This is due to complicated weather patterns, land locations, sea depths, and ocean currents.

It is widely acknowledged that feedback mechanisms are highly influential to the climate. However, some of these mechanisms are as yet unknown, and the magnitudes (and even their signs) are often very much uncertain for the mechanisms that *are* known.

Quick Reference:

- It was stated earlier that the generally accepted global average temperature increase from 1906 to 2005 was 1.1° to 1.3°F. The National Academy of Science (NAS) in 2001 estimated the impact of human-caused CO_2 is about 1.4 W/m² and the impact for *all* human-caused greenhouse gases is about 1.6 to 1.8 W/m². The IPCC AR4, 2007, reports that there is very high confidence that the global average net effect of human activities since 1750 has been one of warming and that this warming can be quantified at +1.6 W/m² (with a range of +0.6 to +2.4 W/m²).

- There are literally dozens of scientific studies that all indicate the length of time CO_2 persists in the atmosphere is between five and ten years. Among these are natural carbon-14 measurements, carbon-14 analyses after nuclear testing, combustion of carbon-14 from fossil fuels, radon-222 testing, solubility of gases in the ocean, and isotopic mass balances (Note, however, that the USEPA states the atmospheric lifetime of CO_2 is fifty to two hundred years).

- The Earth's vegetation is estimated to hold roughly the same amount of carbon as the atmosphere. Earth's soils and dissolved ocean organic matter are both estimated to hold about three times as much carbon as the atmosphere, per Wally Broecker. The United States Geological Survey (USGS) estimates the amount of carbon locked up in projected hydrocarbon reserves (coal, oil, and natural gas) is six to seven times the amount as in the atmosphere. The USGS also estimates there is about twelve times as much carbon (in the form of methane) locked up in gas-hydrates as is in the atmosphere. The oceans are estimated to contain about fifty to sixty times as much dissolved carbon dioxide as the atmosphere. Carbonate rocks contain even more carbon than the ocean. Therefore, it can be seen that the atmosphere contains only a small portion, <1%, of Earth's carbon that influences the atmospheric CO_2 balance.

- The oceanic-atmospheric CO_2 balance is believed by many scientists to be of utmost significance. The possible mechanisms linking this balance to climate are numerous. Cold water can hold more CO_2 than warm water. Warm air can hold more water vapor than cold air. Larger polar ice caps reflect more sunlight into space (ice-albedo). Ocean algae utilize sunlight and CO_2 to generate more algae. Sea animals feed off algae and recreate CO_2. Atmospheric CO_2 is transferred to the ocean because sea animals' fecal matter drops to the ocean floor and shifts the oceanic CO_2 equilibrium with the atmosphere. Plus there are effects from El Niño, freshwater floods, and ocean currents.

- North Atlantic Ocean currents seem in a particularly delicate balance, and it has been suggested that these currents could switch off due to global warming. These ocean currents have been modeled extensively, and the reported results do show sensitivity to increased fresh water additions in the North Atlantic—but this does not link CO_2 to global warming or indicate a *cause* of global warming. There is growing consensus that ocean current switching appears unlikely as a potential *effect* from global warming, as Wally Broecker proposed and Al Gore promoted, because the cause would need to be relatively sudden, such as a large freshwater flood. Plus previous warm periods were warmer and lasted longer than the current warming trend without observing this result.

- Opponents of Global Warming have stated that computer

STEVEN E. SONDERGARD

modeling of surface temperatures indicates that temperature anomalies can be attributed to both greenhouse gases *plus* a solar variation of as much as 0.5%. But this much solar variation is very unlikely. Scientists do know that the solar output from our Sun is not perfectly constant. Solar gain indeed varies about 0.1% over the normal eleven-year sun spot cycle. (This correlates very well with berillium-10 levels—resulting from cosmic rays—in ice core samples.) Longer-term variability in solar output from our Sun also exists. The last several decades appear to be among the most solar active since instrument-based recording began, creating a Modern Solar Maximum characterized by a strong solar magnetic output and increased ultraviolet irradiance that can last for decades or even more than a century, as the Medieval Solar Maximum apparently did—indicated by carbon-14 analysis.

- The amount of sunlight hitting the Earth is ~1,370 watts per square meter (W/m^2). This is the yearly-average, mean-distance, full-spectrum, straight-on solar gain at the top of the Earth's atmosphere. However, due to the curvature of the Earth, the average solar gain at the top of Earth's atmosphere is about 342 W/m^2. The clear-day perpendicular solar impact at the Earth's surface is about 1,000 W/m^2. But this does vary due to atmospheric conditions, the angle of the Sun relative to the surface, and other factors.

- The total solar irradiance (TSI) increased about +0.1% to +0.2% from 1900 to 2000," or 0.2 to 0.5 W/m^2 averaged at the surface. For comparison, the geothermal contribution of global warming has been estimated at +0.06 W/m^2.

- The mathematical correlation from 1875 to 2000 of average CO_2 concentration in Earth's atmosphere versus average arctic surface temperature has a curve fit R^2 of about 0.22.[1] This is not a good correlation.

- Interestingly, the mathematical correlation since 1875 of Hoyt-Schatten TSI versus average arctic surface temperature has a curve fit R^2 of about 0.79.[2] This is a reasonable correlation, but it should also be noted that

Hoyt-Schatten is currently not considered the most representative TSI observation.

- The polar ice caps on Mars are currently experiencing shrinkage, providing us further evidence that our Sun may be increasing in solar output.
- The total impact (including feedback mechanisms) of human-caused atmospheric CO_2 is approximately 0.8°F (~1.4 W/m^2) of warming for the +100 ppm of increased atmospheric CO_2. However, the *direct* radiative warming from CO_2 is believed to be considerably less than that, without including the Earth's multiple feedback and balancing mechanisms.
- The amount of increased warming due to the direct-only increase in TSI is therefore one-sixth to one-third of the estimated direct irradiative climate impact of all human-caused GHGs. Recent calculations and experimentation indicates the total solar effect could be even higher and possibly two to three times the impact of increased TSI—due to increased ultraviolet radiation and cosmic ray impacts.
- Earth's feedback systems are far less understood. But probable increases in solar gain, the CO_2 temperature correlation from ice cores, and today's CO_2 levels and gradients indicate the overall feedback amplification is probably less than a factor of two. Even though some surmise there is a much larger overall natural feedback amplification of the warming from global CO_2 levels, it appears to be unsubstantiated by much of the data. The IPCC estimates an assumed doubling to quadrupling effect based primarily on guesses for two feedback mechanisms: the water-vapor feedback and the ice-albedo feedback. But scientists readily admit that many other undetermined and un-quantified feedback mechanisms could change the magnitude or even the sign of the total feedback effect. In addition, the IPCC appears to have minimized the impact of solar irradiance and dismissed any larger total solar effect as a contributing factor to recent warmer temperatures.

STEVEN E. SONDERGARD

When updated solar impacts are considered, it necessarily lowers the feedback multiple that the IPCC has assumed.

- Scientists know that 1) Earth's orbit is eccentric in nature and the eccentricity varies, 2) Earth's inclination from the orbital plane does vary, and 3) Earth does wobble on its axis. These variances have been calculated, and the evidence indicates that Earth goes through natural temperature variations from these impacts alone. The eccentricity cycle from near circular to more elliptical and back takes about 100,000 years. The tilt variance is from about 21.5° to 24.5° and has a period of about 41,000 years. The wobble cycle is 19,000 or 23,000 years. Interestingly, from ocean sediment core sampling, the oxygen-18 to oxygen-16 ratios of both calcium carbonate and silica, the exact same variance and cycling is observed—determined through Fourier analysis. This indicates that changing solar gain, particularly near the polar latitudes, has caused variations in the polar ice cap masses that have changed the concentration of oxygen-18 to oxygen-16 in the oceans. We therefore have both a rough predictor and a reasonable indicator of the cycling and periodicity of past ice ages going back millions of years.

- The last four major ice ages (over the last 400,000+ years) have been studied from ice-isotopic core samples at central East Antarctica at Vostok. The ice core pattern indicates that the Earth's climate cyclically slides, with wild bounces, into ice ages over about ninety thousand years; then it rapidly warms over about ten thousand years. The most stable climates are realized at the most cold and most warm parts of the cycle. We are currently in an unusually stable and extended warm portion of this cycle. The often-cited Vostok data indicates the CO_2 concentrations of the gas trapped within the ice (from ~180 ppm to ~280 ppm) and the surface temperature variations (from -15°F to +5°F) show a remarkable linear correlation—and through four major ice age cycles. We do not yet know why. Is one the cause and the other the effect, or are they both the effects of something else? Scientists understand that correlation does not imply causation, as many have concluded.

Remember, our atmosphere currently has a CO_2 concentration of 380 ppm, and if CO_2 was indeed the sole cause of the temperature increases, from this trusted data alone, we might extrapolate the data to expect a 16°F to 20°F rise in our average temperature today—which is obviously not the case.

- There is no question that atmospheric CO_2 concentration and temperature are linked. Yet, the best ice core data to date indicates the CO_2 concentration appears to lag behind the average surface temperature by a few centuries. The Vostok data, which is often cited to link temperature increases to CO_2 increases, also shows CO_2 levels lagging atmospheric temperatures. A lag of this nature would be an indicator that the atmospheric CO_2 concentration may indeed be an effect, not a cause, of the temperature increase. This cause-and-effect relationship could be explained in part because the oceans hold significant quantities of dissolved CO_2, and some of that would be driven into the atmosphere when the oceans are heated by increased solar irradiation.

- Ice core samples indicate other consistently periodic temperature oscillations such as the Dansgaard-Oeschger oscillations, spaced generally about 1,500 years apart but with considerable variability. There are ocean iceberg sediment deposits, termed the *Heinrich events*, and their *Bond cycles*. These oscillations and events are explained to be from cyclical Hudson Bay ice formation and subsequent warming. Some suggest we do not yet fully understand their root cause, but others suggest it may be from solar variability.

- Therefore, most scientists agree that at least *some* of the recent warming trend can be attributed to human-caused CO_2. Yet, with the many forces that influence the climate of Earth, it is more difficult to determine *how much* warming is attributed to increased CO_2, and the answer is still of considerable debate. A growing belief is that anthropogenic causes are responsible for a majority of the temperature increases we are observing today. A realistic view, however, is that anthropogenic causes could have a probable impact that might be somewhat less than this growing belief.

STEVEN E. SONDERGARD

More Detail:

There are many factors that appear to impact global warming and climate change. In addition to increasing atmospheric concentrations of CO_2, the other impacts include solar irradiance and solar variability, Earth's orbital variations, ocean currents, and feedback mechanisms. Supporting evidence swaying some opinion includes shrinking polar ice caps on Mars, GHG-temperature correlations, the Dansgaard-Oeschger oscillations, the Heinrich events, and their Bond cycles.

What does all this have to do with how much CO_2 affects global warming? First, it illustrates some of the factors influencing our climate other than CO_2. Second, it provides additional perspective. Third, the discussion provides a general indication of the magnitude of these other factors on our climate. Fourth, any warming or climate change attributed to any of these other factors necessarily means the impact of CO_2 on our observed warming and climate must be less. Conclusions about the impact of CO_2 are included within tables near the end of the chapter.

Solar Irradiance and Solar Variability

It was a surprise to the author to discover that the hundreds of qualified scientists doing research on the solar impacts to climate change are usually looked at unfavorably (sometimes even with disdain) by those who proclaim anthropogenic GHGs are the primary cause of global warming. It appears that any significant findings these solar scientists uncover or any serious alternative theories could be a threat to their position. The IPCC, for one, has at various times minimized and diminished the Sun's influence on Earth's climate. But, whether it is uncomfortable or not, the data should be objectively examined. Changes in sun spots have actually been correlated with weather for more than two centuries.[3]

The solar constant yearly mean-distance straight-on solar irradiation at the top of the Earth's atmosphere is 1,368 watts per square meter (W/m^2), per NASA. The IPCC AR4, 2007, WG-1 states this number as 1,370 W/m². P. J. Michaels and R. C. Balling also state this number as 1,370 W/m².[4] Even though it is called a solar constant, it is not really constant. Every set of solar observations (e.g. Hoyt-Schatten, Lean-Beer-Bradley, Lockwood, Preminger, Solanki-Krivova, Svalgaard, and Wang) shows an eleven-year cycle, and most show an upward trend since 1900—but the lower bounds of the eleven-year cycles are now believed to be flatter. The Hoyt-Schatten total solar irradiance varied from about 1,369 W/m² in 1900, to 1,372 W/m² in 1940, to 1,370 W/m² in 1960, to 1,372 W/m² in 2000.

We observe significant solar variations from many stars all the time, so a small amount of solar variation from our own star should not surprise us. The clear-day perpendicular solar impact at the Earth's surface is about 960 W/m² (per the IPCC) to 1,000+ W/m² (per many others). This number varies more because the energy that hits the Earth depends on upper atmospheric reflections, cloud cover, particulates in the atmosphere, the depth of atmosphere that the sunlight must travel through—a function of both latitude and time of day—and other factors. Under the right atmospheric and seasonal conditions, the value at the surface can even approach 1,200 W/m² for up to four hours around solar noon.

Using simple geometry, if we assume that the Earth is a perfect sphere, which is a reasonable approximation, the surface area is calculated from the formula $4\pi r^2$. The full sphere surface divided by the perpendicular circular area (πr^2) gives us the solar impact over the full surface ($4\pi r^2/\pi r^2 = 4$). Therefore, due to the curvature of the Earth, the average yearly mean-distance solar gain at the top of the atmosphere above the Earth's surface in the year 2000 was about 1,370/4 = 342 W/m². This corroborates the incoming solar radiation found in Figure 2.1. By analogy, the energy from our Sun *at our Earth's surface* is only about 240 W/m² to 250 W/m² averaged over the surface. These numbers will be used in subsequent calculations.

STEVEN E. SONDERGARD

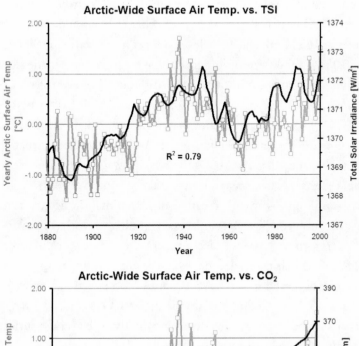

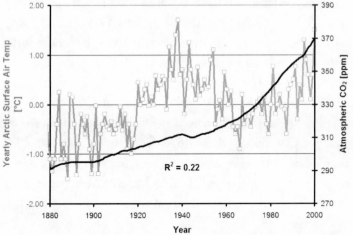

Figure 6.1. Sources: The Arctic Basin wide air temperatures were constructed by Polyakov. The total solar irradiance (TSI) was constructed by Hoyt and Schatten. The correlations were determined by W. H. Soon, 2005. The correlation from 1875 to 2000 of average CO_2 concentration in Earth's atmosphere versus average arctic surface temperature has a curve fit R^2 of about 0.22. The correlation since 1875 of sun spot activity versus average arctic surface temperature has a curve fit R^2 of about 0.79.

The contribution to the twentieth-century warming trend by solar irradiance is about +0.1% to +0.2%, but other effects could more than double the *overall* impact. The eleven-year sun spot variation alone is about 0.1% variability according to satellite measurements made since 1979. For comparison, the Maunder Minimum has been generally estimated at -0.2%, with several estimates as much as -0.4% variability.[5,6,7,8,9,10] There is also evidence that ultraviolet radiation has varied by a few percent, even when TSI was constant. All this information suggests that the +0.1% to +0.2% increase in twentieth-century solar irradiance seems plausible, and it may even be conservative.

In examining Figure 6.1, it can be observed that the atmospheric CO_2 concentration has been increasing rather steadily for over a century. Yet, the average surface temperature shows a decline from 1940 to 1970. This three-decade decline is virtually uncontested and can also be observed in the NASA graphs reproduced as Figure 5.3 and Figure 5.4. A striking question is that if CO_2 is indeed a major cause of surface temperature rise, then why did the average surface temperature decline for roughly three decades when the CO_2 concentration did not? The overwhelming answer is that a temporary force or cause must be even larger than that of CO_2, and it must work in the opposite direction in order to overcome or overpower the impact of CO_2 during these three decades. If that is indeed the case, then what is this larger force? It might even be possible for this larger force to be harnessed and utilized to offset any future CO_2-induced warming. Controlled, atmospheric aerosols might fall into this category. A similar answer is that CO_2 may just not be the largest driving force for the observed changes in average surface temperature. There could be many reasons for this possibility, including negative feedback mechanisms. There is some evidence indicating solar variability might be a larger driving force for the observed changes in average surface temperature, rather than atmospheric CO_2 concentration.

Still another possibility is the explanation forwarded by David

STEVEN E. SONDERGARD

Thompson, 2008, at Colorado State University.[11, 12] Thompson states that the 1940 to 1970 data needs to be filtered due to errant ocean temperature measurements made in canvas buckets during this period by British ships. It is suspected that this method recorded cooler ocean surface temperatures than actual. The number of temperature observations made in this manner by British ships was markedly greater during this period, and Thompson claims the data from this period should be adjusted upward by about 0.2°C. This adjustment would significantly diminish the three-decade temperature decline, but a more important reflection by Thompson's work is the imprecise and inconsistent nature of the composite database being used for global average surface temperature, which serves as the foundation of scientific knowledge concerning climate change assessments.

Lockwood and Stamper, 1999, estimated the impact of solar luminosity accounted for 52% of global temperature change from 1910 to 1960 but only 31% from 1970 to 1999.[13] Scafetta and West, 2006, at Duke University estimated that solar irradiation accounted for 50% of the warming since 1900 and 25% to 35% since 1980.[14] Further, the Sun's eruptional activity, flares, bursts of solar wind, and coronal mass ejections appear to impact global warming. An increase in the Sun's ultraviolet radiation can be absorbed by ozone in the stratosphere, which heats the upper atmosphere and higher latitudes, and then propagate downward into the troposphere.

Evidence exists showing that the total magnetic flux from the Sun has increased by a factor of 2.3 since 1901.[15] The Sun's magnetic field creates a protective umbrella called the heliosphere. Cosmic rays are bent away from the Earth by this heliosphere. Additionally, cosmic rays are thought to ionize atmospheric particles and thereby increase low-elevation cloud formation. This was recently supported by the SKY experiments of Henrik Svensmark, 2006, and his team, who demonstrated at the Danish National Space Center exactly how cosmic rays could increase low-altitude cloud formation.[16] Further experiments are planned at CERN (the European Organization for Nuclear

Research) in a new CLOUD (Cosmics Leaving Outdoor Droplets) facility in Geneva, which is expected to begin operation in 2010.

After including a cosmic ray effect, Shaviv, 2005, estimated the Sun could be responsible for up to 77% of the observed temperature changes of the twentieth century.[17] These sources might indicate that the total solar effect may be a more important driver for global warming than the mainstream assumes. Svensmark, 2007, estimates that low-elevation cloud cover can vary by 2% in five years, and that would affect the Earth's surface by as much as 1.2 W/m^2.[18]

Dr. Habibullo Abdussamatov, head of the Space Research Laboratory in St. Petersburg, believes half of the twentieth-century warming is due directly to the Sun, and most of the other half is due to natural greenhouse impacts and natural variations in albedo.[19] Abdussamatov goes on to state that the Russians will be installing scientific equipment in a space station module by 2010, and this is expected to corroborate the solar influence. The equipment should start collecting data in 2011, and by about 2016 there should be enough data to render a more precise determination.

On the other hand, Lockwood and Fröhlich, 2007, recently found the rapid temperature rise observed since 1985 does not correlate with solar variability.[20] Also, Scafetta and West, 2007, have concluded that most of the temperature rise since 1900 has to be interpreted as human-caused.[21] While it could be argued that atmospheric CO_2 concentration has become the dominant climate driver since about 1985, these conclusions depend on the global temperature database since 1985 being an accurate portrayal of the actual temperature rise. Recall that there are still questions around the integrity and consistency of recent surface weather station data, particularly during this period. This uncertainty further highlights the need for carefully following sound statistical methodology in order to have high confidence in the conclusions.

The recent variability in total solar irradiance, that is a Modern Solar Maximum, might be reasonably estimated as 0.2 to 0.5 W/m^2

STEVEN E. SONDERGARD

averaged at the surface. But the total solar effect may be double or triple this calculated amount due to variability in the *type* of solar output, cosmic rays, and solar eruptional activity. That contrasts with the IPCC AR4, 2007, which stated that the changes in solar irradiance since 1750 are estimated to cause a +0.12 W/m^2 in warming (with a range of +0.06 W/m^2 to +0.30 W/m^2)—which is less than half what was reported in the previous (TAR) IPCC report. Whether ultimately proven right or wrong, it appears that the latest IPCC report has minimized the impact of solar variability on global warming to the lowest of the reported TSI variations. Additionally, the IPCC does not account for any cosmic ray impacts on clouds at all. Neither is a cosmic ray effect included in any of their AOGCM models. The reported Hoyt and Schatten TSI variability does fall within the stated IPCC range of probable solar impact, but just barely.

The National Academy of Science, 2001, estimated that the impact of human-caused CO_2 is about 1.4 W/m^2, and for *all* human-caused greenhouse gases, it is about 1.6 to 1.8 W/m^2. The IPCC AR4, 2007, states there is very high confidence that the global average net effect of human activities since 1750 has been one of warming, with a radiative forcing of +1.6 W/m^2 (with a range of +0.6 to +2.4 W/m^2). *Radiative forcing* is simply a term used by many in the scientific community for the amount of energy from any cause that impacts the climate energy balance. It is usually expressed in watts per square meter, W/m^2.

It can now be concluded that a conservative estimate (without cosmic ray effects) for the portion of warming due to increased solar gain might be 15% to 35% of the estimated radiative forcing of all human-caused CO_2 ($0.2W/m^2/1.4W/m^2$ = 15%, and $0.5W/m^2/1.4W/m^2$ = 35%) Yet, it cannot be dismissed that the portion of warming due to increased solar gain might reasonably be much higher than these percentages. Feedback mechanisms (which may be positive to strongly negative) resulting from these direct global warming mechanisms appear to impact the climate as well.

Shrinking Polar Ice Caps on Mars

Mars's atmosphere is about 95% CO_2, Mars's polar ice caps are mostly made of CO_2 rather than water, and the Martian year is 687 Earth days in length. The polar ice that remains through Mars's summer is called the residual ice cap. There is believed to be some amount of frozen water underneath this residual portion of the CO_2 ice cap. Images of the residual ice cap regions can be compared year to year to determine shrinkage or expansion. Images from the Mars Orbital Camera (MOC) show that Mars's southern polar cap has shrunk a little bit more during each southern summer since the first pictures were taken in 1999.[22, 23] The shrinking of Mars's polar caps was first published in *Science* on December 7, 2001. It is understood that the observational time of Mars isn't very long, but this observed shrinking of Mars's polar caps is supporting evidence for the increasing solar irradiance from the Sun. A common objection to this conclusion is that the observations may be due to natural orbital variations similar to the Milankovitch cycles of Earth (discussed in the next section). The counterargument to this objection is that these natural orbital variations would take millennia, just like on Earth, and the observed changes must be driven by a faster mechanism. Furthermore, there are indications that warming is occurring on the other planets and moons in our solar system that have atmospheres.

Earth's Orbital Variations

It has already been discussed that glaciers are formed from water that has varying levels of oxygen isotopes. As more ice piled up, the ice sheets got bigger, the Earth got colder, the ocean levels got lower, and the remaining oxygen isotope ratio in the oceans changed. It is also known that the ocean sediment layers on the ocean floor contain small shells. These shells are made up of calcium carbonate and silica—both of which contain oxygen. This oxygen comes from the sea water at the time the shells are formed. Therefore, these ocean sediment layers will vary in their oxygen isotope ratios, which in turn reflect the ocean levels and the total amount of ice existing on the planet. The small benthic shells of *foraminifera* were used for the original work, which is attributed to Harold Urey at the University of Chicago, and his protégé Cesare Emiliani. Also contributing was Edouard Bard at Columbia who precisely analyzed the marine core SU81–18 via radiocarbon dating.

Land Ice On Earth

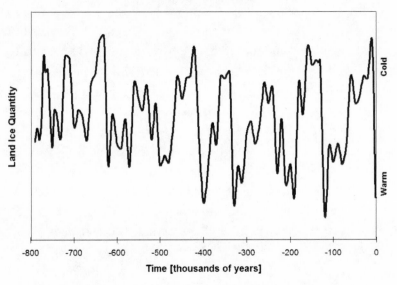

Figure 6.2. Note the large and periodic cycles of ice formation in the past 800,000 years. (Adapted from: *The Two-Mile Time Machine*, Richard B. Alley, 2000, Figure 10.1, p.94)

When the ice quantities are high, more of the lighter isotopes that had more easily vaporized from the oceans are locked up in the ice, and more of the heavier isotopes are in the oceans, which form shells with heavier isotopic ratios. When the ice melts and flows back into the oceans, the oceans regain more of the lighter isotopes, and the shells are formed with the lighter isotopes. (For example: for each 1°C that the ocean water temperature cools, the oxygen-18 is enriched in the shells by ~0.023%.[24]) This mechanism gives us a relatively good historical record of the volume of ice on our planet over millions of years. This work indicates that there have been dozens of ice ages in our past. There are some isotopic variances between specific species of shell creatures, and water temperature also causes a fairly large isotopic shift, but these variances can be overcome and corrected. In 1973, Nicholas Shackleton confirmed the conclusions separately with radioactive potassium measurements in deep sea cores.

What this record tells us is that we have had generally more ice on Earth in the last few million years than existed previously. The ice has formed and melted cyclically many times. We have had major 100,000 year cycles of peak ice formation and reduction over the last million-plus years. There was a repeated spacing of 41,000 years and a spacing pair at 19,000 years and 23,000 years. Because of the variations and fuzziness of the ocean floor data, these period-icities were determined through Fourier analysis. The periodicities also correlate very well with our ice core record. It has even been determined that the 41,000 year oscillation dominated the 100,000 year oscillation prior to a million years ago. Additionally, it can be observed that the last few inter-glacial periods have generally been getting warmer, with less and less land ice. Some have extrapolated this past trend to the current interglacial, which would indicate that further global warming might be expected to occur naturally. It has also been determined that land ice volumes at the peak of the last ice age (the Last Glacial Maximum, or LGM) approximately 20,000 years ago covered about 30% of Earth's land (compared to about 10% today) and the oceans were about four hundred feet lower at the LGM relative to where ocean levels are today.

Each of the periodicities mentioned above were actually hand calculated in the 1910s to the 1930s by Serbian engineer Milutin Milankovitch. These calculations therefore predated by a few decades their discovery in the ocean layers. Milankovitch noted some odd features to Earth's orbit due to orbital influences such as the gravitational tug of Jupiter and Saturn (due to their mass) and even Venus (due to its proximity). He then calculated the periodicities of the sunlight shift that would occur across the Earth. These periodicities do not really vary the *average* amount of sunlight that hits the Earth, but they do impact *where* the sunlight hits the Earth and during what season.

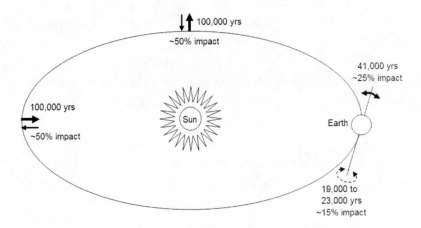

Figure 6.3. Milankovitch's oscillations and their approximate impact on recent ice ages.

Milankovitch calculated that our Earth's orbit is eccentric (it is elliptical rather than circular), that it varies in its eccentricity, that our Earth's inclination from the orbital plane does vary, and that our Earth does wobble on its axis.[25, 26, 27] These impacts are due to the outside gravitational influences of other celestial bodies such as Jupiter, Saturn, and Venus. The eccentricity cycle is relatively complex, but it loosely has an elliptical variation from -0.03 to +0.02 and the cycle takes 96,000 years. The tilt variance is from about 21.5° to 24.5° and has a period of about 41,000 years. The wobble cycle is 19,000 or 23,000 years. The wobbling causes the time when the Earth is closest to the Sun to shift between northern and southern hemisphere. These orbital variations align perfectly with the ocean sediment record of climate variation.

It was a key discovery to realize the periodicities and the timing of ocean sediments matched the Milankovitch periodicities and timing. This was first published by Hays, Imbrie, and Shackleton, 1976.[28] The agreement is far too good to be a coincidence. Even so, scientists do not yet understand the detailed cause-and-effect relationships of these orbital periodicities to our climate. The general connection between the Milankovitch calculations and ice formation can be found in many sources.[29, 30, 31, 32, 33]

Scientists therefore have both a rough predictor (Milankovitch's calculations) and a reasonable indicator (the changes in ocean sediment isotopes) of the cycling and periodicity of past ice ages going back millions of years. The Milankovitch periodicities have even been found in mineral deposits dated 100 million years ago. Small changes in orbital variation and solar radiation are therefore seen to have huge changes in our climate.

A proposed mechanism for a link in the cause of the effect on climate has been forwarded by Rhodes Fairbridge of Columbia University.[34] When two or more of the planets line up on one side of the Sun, not only are the orbits of the planets impacted, but the center of gravity of the solar system is altered. The center of gravity of the solar system could theoretically be as much as one sun diameter outside the Sun. When the center of gravity of the solar system is significantly different than the center of the Sun, it may cause more solar flares, mass ejections, a change in magnetic flux, or sun spot activity. When this idea is coupled with the cosmic ray idea relayed earlier, it can be seen that climate could indeed be impacted significantly with somewhat predictable periodicity.

So what does all this have to do with determining the climate impact of increasing atmospheric CO_2 concentrations? First, it was shown that scientists have reasonably good data on how Earth's climate has varied. Second, there has been extensive climate variability in

STEVEN E. SONDERGARD

the past. Third, any impact that solar variations or orbital oscillations have on climate necessarily means that the climate impact of GHGs are less. Fourth, the fact that water vapor is the most significant GHG means that the climate impact of CO_2 must be even that much less.

The ice core record indicates the glacial to interglacial temperature swings have been about 18°F to 21°F. The Milakovitch oscillations have been attributed to 85% to 90% of that total by matching the ice core temperature record against the Milankovitch calculations. This percentage splits into the orbital eccentricity variability at about 50%, the Earth tilt variance at 20% to 25%, and the Earth wobble at 10% to 15%. These breakdowns are determined by measuring the relative temperature excursions in the ice core record at the corresponding time intervals. The result is that Milankovitch cycles translate to 16°F to 18°F of the glacial to interglacial temperature swings, with eccentricity variation being 9°F to 10°F, Earth tilt variance being 4°F to 5°F, and Earth wobble being 2°F to 3°F.

It could be argued that most of the greenhouse gas impact would be included within that 16°F to 18°F due to feedback mechanism responses. Solar variability might then be attributed to the remainder of about 3°F. If the impact of greenhouse gases is roughly double the magnitude of solar variability, the ice age impact of all greenhouse gases could be about 6°F. The impact of water vapor appears to be over 90% of the total greenhouse impact as indicated earlier. Therefore, a realistic estimate of the impact of ice age CO_2 concentration variability could be 0.6°F to 0.8°F for the impact of the 100 ppm change in ice age atmospheric CO_2 concentration. The 100 ppm increase in *industrial age* atmospheric CO_2 concentration would be expected to have an impact that would be even less, due to the logarithmic effect.

The GHG–Temperature Correlations

It can be observed from our ice core record that the carbon dioxide levels in Earth's atmosphere were relatively low during the peak ice formation of past ice ages and relatively high during the warmer periods of low ice formation. This was first discovered in Bern, Switzerland, and Grenoble, France, at virtually the same time in 1980. Wally Broecker has discussed several of the forces driving this fluctuation.[35] It turns out that there are many explanations, but not one of them is sufficient to explain the full extent of the observed fluctuations in atmospheric CO_2 levels. A complex combination of smaller contributions and offsets involving the ocean is probably the answer.

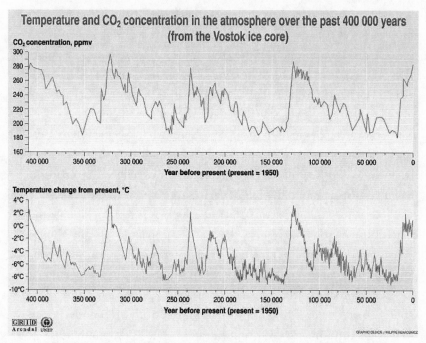

Figure 6.4. The correlation between temperature and CO_2 in the Vostok ice core record. (Source: http://maps.grida.no/go/graphic/temperature-and-c02-concentration-in-the-atmosphere-over-the-past-400000-years; cartographer/designer/author: Philippe Rekacewicz, UNEP/GRID-Arendal; data source: J.R. Petit, J. Jouzel. et. al. Climate and atmospheric history of the past 420,000 years from the Vostok ice core in Antarctica, *Nature* 399 (3 June), pp. 429–436, 1999)

STEVEN E. SONDERGARD

The correlation of the past atmospheric CO_2 levels to atmospheric temperature in the ice core record is actually quite remarkable. Refer to Figure 6.4. There is no question that atmospheric CO_2 concentration and temperature are linked, but there doesn't appear to be a single cause-and-effect relationship. It cannot be concluded that increased atmospheric CO_2 concentration causes an increase in temperature, as is often implied when these ice core graphs of temperature and CO_2 are referenced. There is some data to suggest this, but there is stronger ice core evidence indicating that carbon dioxide lags temperature by four hundred to eight hundred years.[36, 37, 38] The better ice core data for this discernment comes from areas of high snowfall, quick compaction, and low air mixing before bubble close off in the ice—such as Law Dome in Antarctica—but the Vostok data and other widely accepted ice core data indicates this as well. This CO_2 concentration lag to temperature is a very important finding, yet it is either downplayed or ignored by Global Warming proponents. With the ice core record indicating a CO_2 lag relative to temperature, there is a much lower likelihood that atmospheric CO_2 concentration is a primary driver of increased global temperature. One explanation for an increase in temperature causing an increase in atmospheric CO_2 levels is because the oceans hold so much dissolved CO_2 that some of it would be driven into the atmosphere when the oceans are heated by increased solar irradiation or other means.

Methane and nitrous oxide have very similar fluctuations to carbon dioxide, as can be observed in Figure 6.5. The "D" displayed in this figure represents changes in dust levels, not deuterium. The characterization of interglacial periods used here uses a δD exceeding -40.3%. Interglacial periods using this definition are consistent with the marine records. High scattering of N_2O levels is observed with elevated dust concentrations (>300ppbw), and the N_2O record in these ranges are considered disturbed by artifacts and therefore not included in the figure. The atmospheric sources of N_2O are believed to be about two-thirds from soils and one-third from the oceans. The correlation between CH_4 concentration and Antarctic temperature has an R^2 of 0.80.

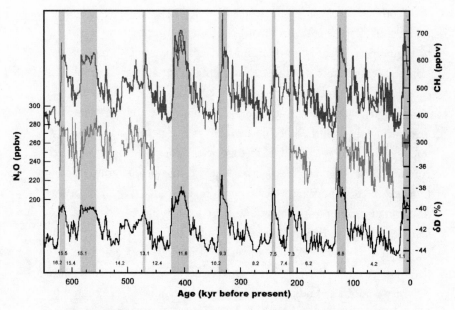

Figure 6.5. CH$_4$, N$_2$O, and dust records over the past 650,000 years; EPICA Dome C CH$_4$ and Vostok CH$_4$ are both included, with the composite record established by wiggle matching. The N$_2$O data is from EPICA Dome C; the dust data is from EPICA Dome C; the gray shaded bars highlight the interglacial periods. (Adapted from: R. Spahni, et al., *Science*, 310, 1317–1321, 25 November 2005)

It is getting common to express the age before present, or year before present, from the year 1950. But be aware that this was/is not always done. It also became a practice in the late 1980s to shift the ice core data forward in time by eighty-three years, after the original Siple, Antarctica, ice core data needed to be shifted to match up with the highly-trusted Mauna Loa atmospheric measurements of CO$_2$. This shift is explained as being necessary due to the air mixing within the ice crystals prior to bubble close off.

STEVEN E. SONDERGARD

The Dansgaard-Oeschger Oscillations and the Heinrich Events and Their Bond Cycles

Over the last 110,000 years or so, there have been periodic, sudden warming phases recorded in ice core layers. These were discovered in the late 1960s to early 1970s by Danish Willi Dansgaard and Swiss Hans Oeschger, in Greenland ice cores.[39, 40] These sudden warmings were roughly 1,500 years apart, with considerable variability. The North Greenland Ice Core Project (NGRIP) recorded twenty-five such Dansgaard-Oeschger oscillations.

Most of the sea floor in the North Atlantic contains at least some iceberg rock debris. However, there are multiple deposition layers that are *predominantly* iceberg rock debris (determined by the chemical makeup of their feldspar and limestone constituents, which match the formations found at Hudson Bay). These were recognized by the German Hartmut Heinrich in the early 1980s, and they are now known as the Heinrich layers, numbered from zero. At least seven Heinrich events have deposited their layers on the ocean floor. These layers can be several feet thick near the Hudson Bay, but they get thinner as you move farther east. They are less than an inch thick in the East Atlantic.

These events and cycles were studied by Gerard Bond from Columbia University, who noticed in 1995 that each successive 1,500-year warming was a little bit milder. After three to five of these successive warming cycles, there would be a significant cold spell after which the North Atlantic would be full of icebergs, depositing iceberg rocks as they melted across the ocean. After the iceberg event, a significant warming would occur, and the cycling would start all over again. This multi-thousand-year cycle with its iceberg events and warming cycles are now known as *Bond cycles*. After further study, Gerard Bond speculated in 1997 that the Heinrich events were caused by a 1,500-year oscillation in solar output.

How do these events and cycles result in such distinct deposits across the Atlantic? It was Doug MacAyeal from the University of Chicago who explained much of the process. As the variability in solar output plus the Earth's 19,000, 23,000, and 41,000 orbital variations both caused solar radiation impacts, there were periods of increased snowfall, and the ice sheets grew through several Dansgaard-Oeschger cycles until the ice sheet eventually filled the Hudson Bay. The ice was then thick enough to insulate and capture the Earth's heat and melt the ice at the Earth-ice interface, forming a slippery mud. The ice then slid quite rapidly over the mud surface, picking up rocks and carrying them into the North Atlantic, creating the Heinrich events. In just a few short decades, the Hudson Bay rid itself of most of the ice that had accumulated over a thousand-plus years. This Hudson Bay ice cycle has a much shorter time period than the overall North American ice sheet, as the North American ice sheet varies on the much slower orbital cycle of approximately 100,000 years.

Tracking the Younger Dryas, which correlates with Heinrich event zero, around the world in ocean and lake sediment was a focus that has provided more clues. The colder parts of these oscillations were also dry and windy over much of the world. The drying would likely occur as the cold and corresponding increase in ice formation reduced the atmospheric water vapor, and that in turn allowed further cooling. The Heinrich events are recorded even in the Southern Hemisphere, although most Southern Hemisphere records show these events as warmer rather than colder.

The roughly 1,500-year Dansgaard-Oeschger cycles may also be tied to ocean currents. The ocean current loop takes approximately 1,000 years to perhaps as much as 1,500 years to complete its entire cycle. The Atlantic currents seem to switch on and off depending on the far northern Atlantic temperature. Extreme cold appears to halt the far northern sinking, causing a large jump from a warmer/wetter/calmer climate to a colder/drier/windier climate. This halting of the ocean currents may explain why the Heinrich events are warmer in the Southern Hemisphere and colder in the North Atlantic, as

STEVEN E. SONDERGARD

the warmed Southern Hemisphere ocean waters would remain in the Southern Hemisphere rather than travel to the North Atlantic.

If the Dansgaard-Oeschger cycles are closely tied to ocean currents, then a puzzle remains as to why the spacing between the 1,500-year cycles has not varied as the climate has changed. Some contend for this reason that this puzzle is yet to be fully understood, and we really don't fully understand the root causes of the Dansgaard-Oeschger oscillations. On the other hand, a roughly 1,500-year oscillation in solar output would need no further explanation. Even with questions about the cause, the bigger picture is much clearer. What is uncontested is that our climate continually goes through significant cycles and changes, and the present period of relative climate stability is quite long when compared with the rest of the climate record. This information illustrates the complex, variable, and sensitive aspects of our climate. Since the Dansgaard-Oeschger oscillations do not correlate with a corresponding oscillation in the atmospheric CO_2 concentration, it further indicates that events other than changing atmospheric CO_2 concentration can dramatically impact climate.

Ocean Currents

The ocean current circuit idea was first developed by Wally Broecker or Arnold Gordon in 1984. Simplistically, the ocean current circuit is described as follows: warm and salty water flows north in the Atlantic near the surface. Then it cools and sinks to flow south on a deeper path in the Atlantic Ocean. The current turns around Africa into both the Indian Ocean and the Pacific where it heats up again. As it heats up it rises to the surface where it is joined by fresh water runoff. This less-dense surface water travels west, back to the Atlantic where it flows north again. The entire circuit takes something like a thousand years. Even though that sounds like a long time, this conveyor system transfers a significant amount of heat away from the equatorial regions and toward the poles. The heat transfer is estimated at about the same magnitude as what is transferred by the atmosphere. The magnitude of the ocean current circuit is roughly equal to that of about one hundred Amazon-sized rivers.

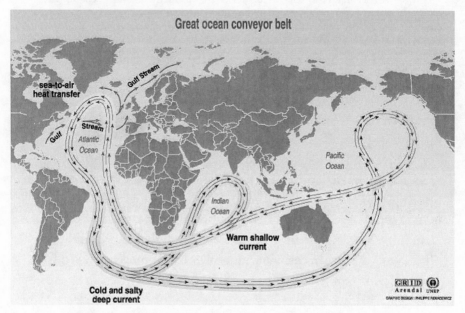

Figure 6.6. The ocean current circulation. (Source: http://maps.grida. no/go/graphic/world-ocean-thermohaline-circulation; cartographer/ designer/author: Philippe Rekacewicz, UNEP/GRID-Arendal; data source: Wally Broecker, 1991, Climate change 1995, Impacts, adaptations, and mitigation of climate change, scientific technical analyses)

This ocean current circuit is affected by water temperature, water salinity, the Coriolis Effect, the location of the continents, rainfall and river runoff, and even wind. The temperature of the water impacts its density, and the warmer the temperature the less dense the water. The salt content also influences its density. Both of these properties together determine whether the water will sink or rise. The fresh water from rainfall and rivers obviously reduce the ocean

STEVEN E. SONDERGARD

salinity as well. The Pacific is generally less salty than the Atlantic, and the northern Pacific is too fresh for much sinking, even though it is relatively cool. The surface water of the Atlantic is more salty than the surface water of either the Pacific, Indian, or Arctic oceans. The equatorial trade winds gather moisture from the Atlantic east of Central America and deposit it in the Pacific west of Central America, which alters the salinity of both oceans.

The Coriolis Effect bends the current flow to the right in the Northern Hemisphere and to the left in the Southern Hemisphere. This effect is due to the curvature and rotation of the Earth. Winds moving into a low pressure center in the Northern Hemisphere will rotate to the left. That is why hurricanes in the Northern Hemisphere always rotate counterclockwise. The northbound currents in the Atlantic bend to the right due to the Coriolis Effect, and they run directly into Europe. They continue their flow north and cool off around Iceland and Greenland. The ocean currents in the Southern Hemisphere bend to the left, but once they pass the tips of Africa and South America, they do not have a continent to lean against or bounce off of. The turbulent Drake Passage between South America and Antarctica is an example of this ocean current circuit. The cold currents approaching Antarctica traverse mile-deep oceans and cool off. They continue eastward along Antarctica where they lose some salinity from melting ice sheets. They are bent northward by the Coriolis Effect towards the Indian oceans and Pacific oceans, where they warm up. The Coriolis Effect continues to bend the currents, and they continue on (westward at this point) to make their way back into the Atlantic again.

This explanation is rather simplified, but you can get a glimpse of the factors that influence the currents moving through the oceans. Three to five million years ago, before the volcanoes of Panama and Costa Rica formed the Isthmus of Panama, the Atlantic Ocean was not separated from the Pacific Ocean. You can now imagine just how much this isthmus must have significantly changed the ocean currents flowing throughout the whole world. Earth's climate in many regions must have been radically changed by this single isthmus formation.

The temperatures in Europe, in particular, are elevated by the warm ocean currents that travel north. Be aware, however, that the perceived magnitude of this long-standing belief in European warming was diminished in 2002 by Richard Seager and David Battisti who measured the heat stored and transported by the Atlantic Ocean and the atmosphere. They then modeled it with two GCMs. They found the winters in Europe are as much as 35°F warmer than North America because the prevailing winds blow from west to east, and blowing across land or water seems to make a huge difference. Further, a standing wave in air circulation is also created by winds blowing over the Rocky Mountains to "bind Labrador" in sea ice. These authors forward that ocean currents therefore add only a "few degrees" of temperature increase for Europe.

One of the things scientists have recently learned about this ocean current circuit is that the North Atlantic turnaround point is a weak link within this global current flow. This region may be so sensitive that if the surface water were a bit less dense, by being either warmer and/or less salty, the current might not sink and start its southward journey. If this were to happen, the entire current circuit could be impacted, at least temporarily. This is believed to have happened before, naturally of course. Computer modeling of the ocean currents indeed indicates that the North Atlantic is a sensitive spot. These models are in their relative infancy, but this modeling can still help us understand what factors might have stopped the ocean currents before and what might stop them again. The models show scientists that sudden surges in fresh water could stop the ocean currents, and that would abruptly impact the climate, similarly to the quick changes observed in ice cores. The rapid release of fresh water from a large, melting ice dam could cause such a current stoppage. Some scientists believe that the sudden failure of an ice dam in the Hudson Bay rapidly dumped a very large amount of cold and icy water into the northern Atlantic, causing the ocean currents to temporarily switch off and trigger the Younger Dryas cold period, which lasted about a thousand years.

It appears from the modeling that once the currents stop it is somewhat difficult for them to start up again. When the currents stop, Greenland and Europe get much colder, and most of the Northern Hemisphere cools. With this cooling the temperature difference between the equator and northern latitudes increases, causing winds to increase, even from as far away as Venezuela. These winds carry dust, which is detected in our ice core records to help confirm this pattern. In addition, the stronger winds cool the surface waters of the ocean, which reduces the greenhouse gas emissions, mostly water vapor, thus cooling the northern regions even more. A colder north appears to impact the warm air movement from Africa and Asia, the monsoons diminish, and these regions become drier. This in turn reduces the global wetlands, CO_2 and methane emissions are thereby reduced, and the planet cools further.

Ocean sediment analysis confirms that when ocean currents have stopped in the past, the planet switched into a colder, drier, windier era. Water from each of the world's oceans has distinctive isotopic and chemical characteristics. When the ocean currents move these waters around the globe and then cease for a time, the composition of the shells of sea life changes just a little. When these shells are deposited over time on the ocean bottoms, we end up with a historical record of how the ocean currents have changed. As you can imagine, this analysis requires painstaking work from many sediment deposits and layers.

If the shells of bottom dwellers are high in cadmium and light isotopic carbon, then they grew in water that had been in the deep ocean for several centuries. Cadmium accumulates in deep ocean water over time as nutrients fall to the ocean floor. Because cadmium can sometimes replace calcium in shells, we have a record of how deep ocean water moved around. This work was originated by Ed

Boyle at MIT. Barium behaves similarly to cadmium in this regard and supports the conclusion. In addition, algae prefer to utilize the light isotope of carbon in its growth at the surface, leaving more of the heavy isotopic carbon in the water. The fecal matter of sea animals that feed on the algae subsequently drops to the ocean floor taking the lighter isotopic carbon with it. By this meticulous analysis, scientists have determined that water has historically sunk in the North Atlantic in two regions—north of Iceland/east of Greenland in the Norwegian Sea and also southwest of Greenland/north of Newfoundland in the Labrador Sea.

Analyses of sea floor records are ongoing, but it appears that the North Atlantic has three modes: the warm mode, where the currents are much like they are today; the cool mode, where water stops sinking east of Greenland; and the cold mode, where water ceases to sink in the northern Atlantic altogether. During cold modes, the water that filled the deep northern Atlantic was in the deep ocean much longer than in the warm modes. The coldest parts of the ice ages and the Dansgaard-Oeschger oscillations correlate with either greatly reduced Atlantic sinking or with the cessation of sinking in the far northern region. The entire stoppage of North Atlantic sinking appears to correlate with the Heinrich events. Data indicates that far northern sinking has stopped for a few centuries in the past millennia, and a pattern is repeated that extends back for a million years or more. When Earth's longer-term orbital variations caused the coldest conditions to occur, the far northern sinking appears to have stopped for longer periods of time, perhaps millennia.

These ocean current changes have cyclically impacted the weather patterns in the Northern Hemisphere, but they do not appear to have impacted the Southern Hemisphere as significantly. Even so, the northern climate and the plants and animals that live in the Northern Hemisphere have been dramatically impacted in the past by repeated climate changes. This subject is described further by Peter Ward, 2007, and it is touched upon later in Chapter 8.[41]

What triggers these repeated events, outside of the orbital variations, is still uncertain. But some evidence seems to point us to freshwater flooding as a probable culprit. A huge ice-melt-water flood occurred down the St. Lawrence River valley just prior to the Younger Dryas cold surge. Another melt-water flood also appears to have occurred about 8,200 years ago, corresponding to the period of sharply colder temperatures—about 10°F colder in central Greenland—that started about 8,400 years ago and lasted for about 180 years.

There may still be uncertainty around the root cause of the Dansgaard-Oeschger oscillations, as mentioned earlier. But the approximately 1,500-year spacing of the Dansgaard-Oeschger oscillations appear to have similar periodicity to the roughly millennial cycle of the ocean currents, which leads climatologists to speculate there may be a connection. Some speculate the root cause may be due to a 1,500-year variation in solar gain that occurs for some reason repeatedly and periodically. Others speculate it may be due to El Niño or even millennial changes in the Antarctic. What is soon realized, however, is that the North Atlantic Ocean current is in a delicate balance, and small changes can have large climate change impacts that can last for centuries.

Some, including Al Gore, have extrapolated this theory to suggest the ocean currents might be switched off as a potential effect from Global Warming. The result becomes a paradoxical flip-flop in which global warming causes a major cold spell. However, this outcome appears unlikely because the impact would need to be relatively sudden, such as past catastrophic ice dam collapses resulting in sudden ice-melt-flooding. In addition, previous warm periods (e.g. the Medieval Warm Period) were warmer and lasted longer than the current warming trend without observing this result. Furthermore, the concept is fundamentally flawed in that the temperature would need to rise by 7°F to 9°F in order to flip the switch, and then the subsequent cold spell would not create temperatures much colder than prior to the temperature rise. Even Wally Broecker himself, who originally forwarded the idea, has more recently backed way off from, and even apologized for, his initial alarmist position.

There are several central mechanisms that have been proposed for global climate changes. One mechanism comes from the Wally Broecker camp, and it is based on North Atlantic Ocean conveyer circuits where greenhouse-induced warming of northern glaciers and the increase in transport of tropical water vapor to high latitudes increases the freshwater runoff to overload the North Atlantic currents and throws a switch to shut down the ocean conveyer triggering a drastic cooling such as the Younger Dryas. But other camps of investigators believe that other causes are the epicenter for determining much of our past climate changes. For example, Lonnie Thompson proposes a mechanism in which orbital precession causes seasonal changes in tropical insolation. This starts the interstadials and heats the tropical air, which takes on more moisture, causing growth of wetlands and causing methane to increase, and the combined moisture and methane causes a greenhouse effect to heat the tropics and subtropics. A third mechanism for both orbital and millennial climate cycles is proposed by Mark Cane in which the Indonesian warm pool near the Makassar Strait triggers a heating of the Pacific Ocean, causing a perpetual El Niño, and launches the interglacial warmings. A fourth mechanism is based on the correlation between the changes in the tropics and the Antarctic. This Southern Ocean camp is promoted by those who have drilled Antarctic ice cores or Southern Ocean mud.

It should not be too surprising to learn that the location of one's belief about the epicenter of climate change matches up with one's center of research effort. This causal stance is critical to keep one's research money flowing. In reality, it will probably turn out to be a combination of causes, or maybe that different causes might be responsible at different times. Research is still ongoing, but it should be clear at this point that a number of mechanisms other than CO_2 could drive global warming or trigger a change in the climate. This

STEVEN E. SONDERGARD

point is virtually uncontested. It should also be clear that orbital variations and solar variations both appear to have large influences on Earth's climate. It is the relative contributions of these aspects that are usually contested. Even with the evidence presented, Global Warming proponents contend that increasing atmospheric CO_2 concentrations could still be significantly warming our planet, this warming could have dire consequences, the warming might trigger an unstable climate mechanism, and it would be better to avoid these risks even if the cost is high. These risks will be addressed shortly.

Feedback Mechanisms

Feedback mechanisms are the many processes that either amplify or diminish the direct radiative forcing of greenhouse gases (and other natural climate drivers) and thereby impact the climate response. It can certainly be observed in our climate record that over long time periods our Earth's feedback mechanisms act to oppose complete runaway climate changes. Therefore, the combined long-term feedback mechanisms work to diminish the direct impact of global warming. There is virtually no question about this conclusion, as we have many examples of this occurring in our climate record. But there is also widespread belief that over shorter time periods the feedback mechanisms amplify the direct irradiative impact of greenhouse gases. One of the supporting reasons for this belief is the observation that past short-term climate changes are of the same magnitude as long-term climate changes. Several of the many feedback mechanisms are discussed below:

> Albedo—albedo is simply the measure of Earth's reflectivity, and it is defined as the fraction of solar reflection relative to the total solar radiation impacting a body. Surface albedo in the IPCC AR4 AOGCMs was modeled at 0.26 ± 0.08 W/m²/°C.

Ice albedo—this is the albedo due to surface ice. Note that as the Earth warms it would reduce the quantity of surface ice. As surface ice is reduced, less electromagnetic radiation is reflected into space, and the Earth is warmed even more. This feedback mechanism therefore would amplify the impact of direct radiative warming. Both glaciation and deglaciation will amplify the climate trend in this manner.

Cloud albedo—this is the albedo due to clouds. As the Earth warms, the atmospheric humidity will likely increase as well from warmer oceans. As the humidity increases, more clouds should result. With more dense, low altitude clouds, there will be more reflection back into space during the day, but the increased water vapor will retain more of the infrared radiation during the night. This will result in cooler daytime temperatures but warmer nighttime temperatures. If the clouds formed are of the sparser, high altitude variety, then both day and night warming can be expected. The overall impact of cloud albedo is therefore believed to result in a more temperate climate. Cloud feedback in the IPCC AR4 AOGCMs was modeled at 0.69 ± 0.38 W/m^2/°C. There is a relatively large uncertainty in this feedback and in how to model it.

Water vapor feedback—the obvious impact of this feedback mechanism is that as the atmosphere warms the oceans will also increase in temperature, and the result is that more water will vaporize into the atmosphere. This increased water vapor will in turn increase the greenhouse effect and further warm the environment. Water vapor changes are stated by the IPCC to be the largest feedback affecting climate sensitivity, with about a 50% positive feedback. What is less understood is the longer-term impact on clouds, wind, precipitation, CO_2 absorption back out of the atmosphere due to precipitation, and other potential feedback aspects of increased water vapor. Cloud feedbacks remain the largest source of uncertainty, per the IPCC. There are still questions around whether an increased CO_2 concentration impacts the greenhouse effect of water

STEVEN E. SONDERGARD

vapor or not. Because the average water vapor concentration in the atmosphere is so significant, small changes of only a few relative percentage points in this average concentration can have enormous multiplying impacts on the entire global warming picture. Water vapor feedback in the IPCC AR4 AOGCMs was modeled at 1.8 ± 0.18 W/m²/°C.

Humidity lapse rate—this is the change in humidity with altitude, with the humidity dropping significantly at higher altitudes. This feedback is sometimes combined with the water vapor feedback. The humidity lapse rate feedback in the IPCC AR4 AOGCMs was modeled at (negative) -0.84 ± 0.26 W/m²/°C.

Cloud coverage variability—Roy Spencer, 2007, discovered from satellite data that the greenhouse effect of cloud variability behaves with an apparently strong negative feedback.[42, 43] As the tropical atmosphere warms from rainfall, heat-trapping clouds shrink, allowing more heat to escape into space, creating a cooling effect. Current mainstream theory and current modeling does not consider this negative feedback effect.

Carbon dioxide balance—the ocean-atmospheric balance is quite significant. The oceans hold about fifty times as much carbon dioxide as the atmosphere. When the oceans are heated by whatever means, some of the dissolved CO_2 is transferred from the oceans to the atmosphere. Therefore, the increased carbon dioxide from ocean heating could further increase warming and thereby amplify the increased greenhouse gas impact.

Interference greenhouse effect—it is widely acknowledged that the interference between GHGs can reduce the overall greenhouse impact of a specific GHG gas by one-quarter to nine-tenths of its projected single-concentration impact. The interference effect depends greatly on the overlap of the absorption bands for specific GHGs. Specifically, the overlap of the absorption bands for water vapor and CO_2 is significant at roughly 25%.

Release of methane—the cyclical methane concentration in the atmosphere with past ice formation is believed to be due to at least four mechanisms. First, as the Earth gets warmer, the amount of vegetation should increase, and the sources of methane production are increased as well. These sources include thawing permafrost, peat deposits, and wetlands. Second, as the Earth gets colder and ice forms over more of the surface, the methane production from below the surface produced from the mechanisms of biogenic activity, biodegradation of oil, and Archea generation is essentially capped off and is not allowed to escape into the atmosphere. Conversely, this methane would be released in greater quantities after the ice melts. Third, methane-ice hydrates would emit their methane when they melt. Fourth, thermokarst lakes, newly formed lakes from glacial thawing, contribute to methane production. Each of these feedbacks would amplify the direct impacts of global warming and cooling.

Cosmic rays help form low-altitude clouds—cosmic rays are believed to increase low-altitude cloud formation. This was recently supported by the experiments of Svensmark, 2006, who demonstrated exactly how cosmic rays could make dense, low-altitude clouds.[44] The low-altitude clouds increase the cloud albedo and diminish the overall impact of global warming. The combined impact could be quite large, as Shaviv, 2005, estimated the Sun could be responsible for up to 77% of the observed temperature changes in the twentieth century.

Weakening of carbon sinks—it is believed that overall carbon sinks may be weakening in their ability to absorb carbon dioxide. The main carbon sinks are the oceans, soils, and plants. If the ocean gets warmer, it should release more of its dissolved CO_2 and further increase the greenhouse effect. There is also a concern that if the ocean gets more acidic from higher levels of CO_2, a reduction in plankton could

STEVEN E. SONDERGARD

result, which would affect the amount of CO_2 uptake. As deforestation occurs, it should reduce the CO_2 absorption from the atmosphere. This would be further exacerbated if weather patterns result in increased droughts and reduced quantities of living plants. An article by Le Quéré et al., 2007, states that oceans and land absorb half of the human-caused CO_2 emissions and suggests that the Southern Ocean sink of CO_2 has weakened in the last twenty-five years due to increased Southern Ocean winds and surface air temperature.[45]

Delayed impact—the delayed impact is due in part to the ability of the Earth to assimilate increased quantities of greenhouse gases, predominantly CO_2. When the quantity of CO_2 being produced surpasses the quantity that can naturally be assimilated by whatever mechanisms, the quantity of CO_2 in the atmosphere will increase. This appears to be what is happening today. Therefore, the Earth is effectively trying to catch up with the present emission levels being produced. But since it cannot, the atmospheric levels of CO_2 keep increasing. In this scenario, even without an increase in the yearly emission levels of CO_2, the concentration of CO_2 in the atmosphere will continue to increase for at least some period of time. This is due in large part to the longer atmospheric lifetimes of greenhouse gases versus water vapor and the long circuit time of the ocean currents. With this delayed impact mechanism and the present trend of increasing annual emission levels, a stabilization of atmospheric GHG levels will be even more difficult to achieve.

All the feedback mechanisms combined with the direct warming effects are often described with the term *climate sensitivity*. Climate sensitivity is defined as the global mean surface temperature change due to a doubling of the atmospheric concentration of CO_2. The climate sensitivity per the IPCC AOGCMs is 3.2°C with a range of 2.1°C to 4.4°C (5.7°F with a range of 3.7°F to 7.9°F). The climate sensitivities of each of the twenty-three IPCC AOGCM modelings that provide us with this answer are listed separately in chapter eight of the AR4 WG-1 report.[46]

Determining an Overall Feedback Multiplier from Anthropogenic Causes

The direct irradiative warming is often calculated from the pure physics of radiation by first taking the derivative of the Stefan-Boltzmann law (the law of physics describing the relationship between thermal radiation and temperature). After plugging in observed temperature increases, the resulting feedback factor is $0.3°Cm^2/W$. The radiative forcing due to a doubling is $3.75 \ W/m^2$. When these are multiplied together, it yields a direct irradiative temperature increase of $1.12°C$ (~$2°F$) for a doubling of CO_2 concentration. The overall feedback multiplier per the IPCC AOGCMs thus becomes about 2.9, with a range of 1.9 to 4.0. There are several arguments that have been raised with this calculation, however.

The first argument is that unless the observed temperature increases are laboratory derived, they would already include natural feedback mechanisms. Thus, when they are substituted into the derivative of the Stefan-Boltzmann law, the resulting irradiative warming would therefore include feedback effects as well. Obviously, the feedback effects should not be multiplied twice. A second argument raised is that the natural observed temperatures would necessarily include the effect of solar variability, and this would result in a higher direct temperature increase for the greenhouse effect than what is actually the case. A third argument is that this calculation deals only with the radiative aspects and does not consider possible convective, boundary layer, and conductive impacts. Considering these issues, the overall feedback multipliers used by the IPCC AOGCMs and their subsequent climate sensitivities could be high by a factor of about two to three.

STEVEN E. SONDERGARD

It should also be noted that IPCC feedback multipliers are determined by the assumptions and chosen correlations used in their AOGCM models. As such, the overall multiplier can be more the result of the modeler rather than the applied physics and thermodynamics from the model itself. The magnitude of solar variation can therefore be thought of more as an *input* within these models. The physics and thermodynamics of the models do not need to be wrong for the feedback multipliers to be wrong. Until better information clarifies the total solar effect, a realistic overall feedback multiplier is forwarded to be around 1.7, with the possibility of a 30% to 50% increase in the feedback multiplier per doubling of CO_2.

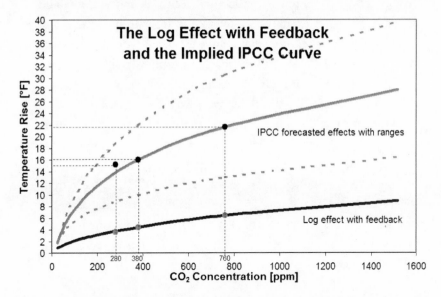

Figure 6.7. The log effect with feedback curve (shown earlier) is shown again here relative to an implied IPCC curve. This IPCC log curve implies that the contribution of CO_2 to the total greenhouse effect is currently about 25%, which is three to five times greater than what others in the literature have concluded. Note that a key pre-industrial data point at 280 ppm CO_2 does not lie on the IPCC curve, but it does lie on the log effect with feedback curve forwarded here.

Another methodology may be the most glaring evidence for a lower overall feedback multiplier. Figure 6.7 shows the past 100 ppm increase in atmospheric CO_2 concentration along with the corresponding observed temperature increase. Note that the pre-industrial data point at 280 ppm could not lie on *any* logarithmic curve going through the IPCC climate sensitivity forecast and the current data point at 380 ppm. With the climate sensitivity slope forecasted by the IPCC being larger than what has actually been observed over the last century, the projected IPCC curve would need to be exponential in shape. For this to be correct, feedback mechanisms must become strongly exponential—in order to overcome the natural logarithmic effect. A strongly exponential feedback multiplier would obviously be extreme; it is counter to the logarithmic concept of climate sensitivity, and it does not appear substantiated by the data. Some increase in the overall feedback multiplier could be reasonably expected, but a radically increasing feedback multiplier may be more than a balanced and realistic view would suggest. It must also be conceded by IPCC proponents that an exponential curve infers that current CO_2 levels would have a minimal temperature increase on our world, because the curve must go through the origin as well. These are both unlikely, and the forecasted IPCC feedback multipliers therefore appear too large relative to known data. Even on the chance that feedback mechanisms are truly that exponential, our Earth's climate would then be so unstable that natural causes alone would be expected to soon trigger a temperature excursion anyway.

A Reasonable Assessment of Greenhouse Gas Warming Due to CO_2

The total warming impact of human-caused CO_2 can be calculated by starting with the 35°C of total overall greenhouse gas impact that was mentioned in Chapter 2. (Note that the 35°C is a reasonable middle-of-the-road number within the 31°C–40°C range of the

STEVEN E. SONDERGARD

various references.) The 35°C equates to 63°F. Next, the contribution of CO_2 towards greenhouse warming as mentioned previously is between 2% and 8% (a realistic number could be as high as 7%). Therefore, the greenhouse gas impact for CO_2 may be about 7% of 63°F, or 4.4°F, with a range of 1.1°F to 5.8°F (2% of 31°C to 8% of 40°C)—for the entire 380 ppm of CO_2 in the atmosphere. This result is slightly higher than the empirical 2.5°F (about 4%) obtained at the University of Delaware, the University of Virginia, and Arizona State University. An increase of 2.5°F was also obtained by P. J. Michaels and R. C. Balling using the NCAR Community Climate System Model 3. This information suggests the 7% contribution for CO_2 may be generous.

Assuming a linear relationship, which is a lenient assumption that some promote, the last 100 ppm of atmospheric CO_2 concentration would have increased the atmospheric temperature about 1.1°F (with a range of 0.3°F–1.5°F). If a more appropriate logarithmic relationship is used, the result is ~0.8°F (with a range of about 0.2°F–1.0°F) for the impact of human-caused CO_2 in the twentieth century. This equates to ~0.9°F for the impact of *all* anthropogenic greenhouse gases. These temperature increases inherently include all feedback mechanisms.

A crosscheck by another means would be in order. The Tropic of Cancer and the Tropic of Capricorn are currently about 47° apart. Therefore, the peak sun azimuth changes about 47° between summer and winter at latitudes between the Tropic of Cancer and the Arctic Circle. Using trigonometry, the peak solar gain per square meter changes annually by about 32% of 342 W/m², or 109 W/m². This annual variation in solar intensity corresponds with an annual swing in observed average inland temperatures of about 45°F. That is ~45°F per ~109 W/m² of radiative forcing, or 0.41°F per 1 W/m², and this includes all feedbacks. Using this same temperature sensitivity to radiative forcing along with the recent increase of 100 ppm of CO_2 having a radiative forcing of 1.4 W/m², per the IPCC and

NAS, it would be expected that the recent increase of 100 ppm of CO_2 would have resulted in a temperature increase of **~0.58°F** for the twentieth century. This result is somewhat less than the previous result, but integrating the summer and winter solar intensity across the sky yields a more accurate result that is closer to 0.8°F.

Table 6.8 Reasonable Factors and Effects:

Aspect	Temp Rise	Forcing Factor	Percent
Anthropogenic CO_2 (+100 ppm)	≤0.8°F	1.4 W/m²	66% or less
Anthropogenic Other GHGs	≤0.1°F	0.2 W/m²	10% or less
Anthropogenic Total	≤0.9°F	1.6 W/m²	76% or less
20th Century Solar Impact	≥0.3°F	0.5 W/m²	24% or more
Natural Water Vapor Response	???	unquantified	unknown
Overall 20th Century Observed	~1.3°F		100%
Direct Doubling of CO_2	~1.5–2°F		
Doubling of CO_2 with Feedbacks	~2.5–4°F (3.7–7.9°F per IPCC)		
Overall CO_2 (380 ppm)	4°–8°F		
Water Vapor Impact	54°–59°F		
Total Greenhouse Effect	60°–63°F		
Dense Cloud Albedo		50–75 W/m²	
Annual ~47° Sun Azimuth Swing	~35°F–50°F	~100–150 W/m²	
Total Incoming Solar (average)		342 W/m²	

Actually, this climate data holds together rather well with a reasonable logarithmic effect, a reasonable overall feedback multiplier, and conservative solar variations. Yet, what about the unknown response by water vapor and clouds? This impact could be quite significant, and it has been projected to have a strong negative feedback. What if the cosmic ray impact on clouds becomes fully established and widely accepted? If an increased total solar effect is later confirmed to be double or possibly triple the total solar irradiance (a rough estimation), the temperature increases associated with each contributing aspect would need some redistribution from what was just presented. With rough adjustments for the cosmic ray effect, the overall impact of human-caused CO_2 would then have contributed *at most* 0.6°F, the total solar effect on global warming would have instead been *at least* 0.6°F, all other human-caused

STEVEN E. SONDERGARD

greenhouse gases would have contributed less than 0.1°F in warming, and these numbers do not include any blind contribution for the projected strong negative feedbacks for water vapor and cloud responses.

Table 6.9 With Consideration for Cosmic Ray Effects:

Aspect	Temp Rise	Forcing Factor	Percent
Anthropogenic CO_2 (+100 ppm)	≤0.6°F	1.4 W/m²	45% or less
Anthropogenic Other GHGs	≤0.1°F	0.2 W/m²	7% or less
Anthropogenic Total	≤0.7°F	1.6 W/m²	52% or less
100yr Solar Impact (+ cosmic rays)	≥0.6°F	≥1.4 W/m²	48% or more
Natural Water Vapor Response	???	unquantified	unknown
Overall 20th Century Observed	~1.3°F		100%
Direct Doubling of CO_2	~1–1.5°F	1.5–2.8 W/m²	
Doubling of CO_2 with Feedbacks	~1.5–3°F	2.8–5.6 W/m²	
Overall CO_2 (380 ppm)	3°–7°F		
Water Vapor Impact	54°–59°F		
Total Greenhouse Effect	60°–63°F		
Dense Cloud Albedo		50–75 W/m²	
Annual ~47° Sun Azimuth Swing	~35°F–50°F	~100–150 W/m²	
Total Incoming Solar (average)		342 W/m²	

The data that appear to be out of kilter are the current AOGCM computer models that indicate much higher temperature sensitivity to CO_2 concentration. These AOGCM models do not appear to properly account for the impact of solar energy or solar variability. Clouds and aerosols are modeled insufficiently. They also do not account for known CO_2 gradients, and they appear to overstate the probable impact of feedback mechanisms. Therefore, it is the AOGCMs that stand out as most questionable. Unfortunately, the IPCC places most of their trust in and performs most of their forecasting based upon these AOGCM models.

It should be apparent, but it will be clearly stated, that these tables include calculations that are *very rough estimates* and that have relatively wide probable ranges. The tables are presented for the reader to get a feel for how the data compares in aggregate. In reality, there are many feedback mechanisms, and their quantifications are still poorly known or even completely unknown. The natural water vapor variability is shown as a placeholder for potentially several natural factors that contribute to global warming and that are simply unknown or unquantified. The quantification of the total solar effect is also still very much uncertain, as was previously indicated.

STEVEN E. SONDERGARD

7. WHAT IMPACT DOES MANKIND HAVE ON INCREASED ATMOSPHERIC GREENHOUSE GASES?

Focus: Multiple producers of greenhouse gases are presented for perspective: the Archaea domain, the potential of living plants (both sides of the argument are presented), human and animal respiration, agriculture and livestock, and human-caused CO_2-equivalent emissions (broken down by category). Then the overall anthropogenic impact per the IPCC is discussed.

Layman's Brief:

Many convey that GHGs are the heart of the global warming problem and fossil fuel burning by man is the single cause, with Global Warming being the consequence. On the other hand, many natural causes impact the atmospheric CO_2 concentration. These natural causes include ocean temperatures, volcanic eruptions, respiration from all animals, and the consumption of CO_2 by plants. The natural emissions of other greenhouse gases can also be significant. One example is the single-celled Archaea domain that generates large quantities of methane below the surface. This methane escapes into the atmosphere naturally. It has also been forwarded that living plants might contribute substantial methane emissions.

Water vapor has been shown to be the most significant greenhouse gas in the atmosphere. Yet, it is readily apparent that the atmospheric concentration of water vapor is almost exclusively natural in its origin. Water vapor can also be impacted by climate feedback mechanisms. But it is still widely accepted that mankind has very little direct impact on the atmospheric water vapor concentration.

Human activities are believed to be increasing the concentrations of greenhouse gases (mainly CO_2) significantly in the atmosphere. Approximately 30% of the atmospheric concentration of CO_2 is attributed to the impact of man. It is calculated that man is currently contributing CO_2 to the atmosphere at about one-thirtieth of the entire atmospheric CO_2 quantity every year. It is acknowledged that less than 40% of this global quantity is contributed by industry. In the U.S., industry contributes less than 30%. It has also been calculated that about half of the CO_2 emissions are assimilated by plants, soil, and the oceans.

The concentration of CO_2 has varied naturally over the last several hundred thousand years between 180 ppm and 280 ppm—until the last couple of hundred years. The concentration of CO_2 in the atmosphere is now about 380 ppm. The increase from 280 ppm to 380 ppm is attributed to man. There is additional evidence indicating that over the past 600 million years the atmospheric CO_2 concentrations generally varied between a low of about 180 ppm to over 5,000 ppm. The average CO_2 concentration over this historical period is believed to be just under 2,000 ppm. It should be obvious that these higher prehistoric levels in concentration were natural in origin.

Quick Reference:

- For historical perspective there are a variety of measurements indicating that significantly more CO_2 has existed in the Earth's atmosphere in the past than exists today (generally from 1,000 ppm to 5,000 ppm prior to 50 million years ago—per NAS 2001). Both the GEOCARB GCM model of prehistoric CO_2 concentrations and the measurement of stomata on fossilized leaves support these high prehistoric levels.

- Also for historical perspective: CO_2, CH_4, and N_2O have each gone through significant fluctuations between past ice ages that can be considered natural. Even so, current atmospheric CO_2 levels are rising faster than at any time in the past that we are aware of, and atmospheric CO_2 concentrations are currently greater than during any past ice age cycle yet measured.

- There is currently an estimated 750–800 billion tons of CO_2 in our Earth's atmosphere. Mankind is being blamed for about 30% of that. This correlates to an increase in atmospheric concentration from about 280 ppm to about 380 ppm.

- Several sources have estimated yearly anthropogenic emissions. The U.S. Department of Energy, National Energy Technology Laboratory (USDOE/NETL), estimates 24 billion tons/year of CO_2 were emitted globally in 2004 (the latest year for which global information is available). The U.S. Department of Energy, Energy Information Administration (USDOE/EIA), estimates 28 billion tons/year of CO_2 were added globally in 2004, with 29 billion tons/year in 2005. The IPCC estimates 30 billion tons/year of CO_2 were emitted globally in 2004. These estimates would double the entire global CO_2 level in twenty-seven to thirty-three years without some kind of natural reduction mechanism(s). The main natural reduction mechanisms are the oceans and land, which are believed to currently soak up roughly half of these anthropogenic emissions.

- Still, it is the *incremental* increase of all greenhouse gases that is of current concern (and it is harder to determine the impact—particularly if the interference with, and secondary impact to, water vapor is considered).
- It is estimated that nature generates from thirty to fifty times as much greenhouse gases as does man (most of it as water vapor). But all of the greenhouse gases that man emits are believed to have a much longer-term warming impact in the atmosphere than water vapor. This is the basis for measuring greenhouse gases by their one-hundred-year GWP.
- The largest contributor of greenhouse gases—by far—is due to the heating of the oceans by the Sun, which causes water vapor to emerge and form clouds (and it even causes CO_2 to emerge). But the water vapor content in the atmosphere varies significantly and often—due to saturation (causing rain and snow), its lower vapor density (causing lower pressure zones and wind), and the dual ability of clouds to cause both warming and cooling: 1) high-altitude, thin clouds cause both daytime and nighttime warming, and 2) lower altitude dense clouds cause daytime cooling with nighttime warming. It is therefore extremely difficult to determine the longer-term influence of water vapor on climate change or how it acts as a long-term feedback mechanism.
- It appears likely that the increase in atmospheric CO_2 concentrations can be attributed (at least partially) to increased ocean temperatures due to increased solar irradiance. The basis for this belief is that increasing solar irradiance appears to be increasing ocean temperatures, and that would drive more dissolved CO_2 from the oceans into the atmosphere.
- The Stern Report states the global anthropogenic CO_2-equivalent emissions in year 2000 were 42 billion metric tons. Of this quantity, the power sector contributed 24%, land use contributed 18%, the industrial sector contributed 14%, the transportation sector contributed 14%, agriculture

STEVEN E. SONDERGARD

contributed 14%, buildings contributed 8%, other energy sources contributed 5%, and waste contributed 3%.

- Note that global industry contributes less than 40% of total man-made global GHG emissions (per the Stern Report).
- Also of note is that 40% of the global CO_2 emissions are attributed to the U.S. and China, with China estimated to have just surpassed (2007) the U.S. because of new coal-fired power plants. Indonesia is gaining fast and is approaching third place due to deforestation.
- In the U.S., for 2005, the USEPA stated that 8 billion tons of human-caused CO_2-equivalent greenhouse gas emissions were emitted. Residential accounted for 17%, transportation accounted for 28%, commercial buildings accounted for 17%, agricultural accounted for 9%, and industry accounted for 28% (2.25 billion tons—which has been on steady decline for more than a decade). An estimated 1.5 billion tons of CO_2 came from U.S. coal-fired power plants. Specifically excluded were consideration of water vapor, human and animal respiration, military activity, and forest fires (human-caused or otherwise).
- It is worth highlighting that industry within the U.S. contributes less than 30% of total human-caused U.S. GHG emissions (per the USEPA).
- Plate tectonics (impacting natural volcanic eruptions) combined with the rate of weak carbonic acid rainfall and rock weathering impact the Earth's CO_2–silicate/carbonate cycle. This cycle is very long, at maybe 100 million years. In an average year, volcanoes emit roughly 0.3 to 1.0 billion tons of CO_2. However, relatively sudden spikes do occur in this cycle from volcanic eruptions, which can suddenly add multiple atmospheric pollutants, CO_2, and particulates and thereby impact the climate (for example, the twelve months following the 1815 Mount Tambora eruption is called "the year without a summer").
- Consider also the CO_2 and methane emissions from the bodily functions of all of Earth's animals and the gaseous emissions from all microbes (including the substantial

Archaea domain beneath Earth's surface). These are not insignificant collective CO_2-eq quantities relative to mankind's contribution through industrial sources.

- In January 2006, *Nature* reported the findings at Max Planck Institute that living plants contribute 10% to 30% of the methane entering the atmosphere (63 to 236 million metric tons per year, with plant debris adding another 1 to 7 million metric tons per year). If this finding holds up, it leads us to question the quantification of other methane sources in the methane balance; plus, it has implications on climate policy that would promote the use of plants (forestation) as CO_2 offsets and sinks.

- The 2001 IPCC report (TAR) stated a "66% to 90% chance that human activities are driving recent warming." The 2007 IPCC report (AR4) stated global warming is "very likely" caused by mankind and that climate change will continue for centuries even with reduced greenhouse gas emissions.

- Note these stated human activities include many aspects that may not necessarily be considered bad (at least they would usually be considered to have *some* degree of positive value). In some semblance of decreasing value, these aspects include human respiration, emissions from livestock raised for food, deforestation for farming and habitation, electricity generation, heating/cooling of buildings, transportation vehicles, and stationary industrial sources that provide jobs and improve the quality of life. Reducing these aspects (if required) would be difficult for many people to accept.

- Interestingly, global human respiration (breathing) emits between 5% and 10% of all human-caused CO_2. Livestock plus pet respiration may be an additional CO_2 contributor of similar magnitude. Further, the UN Food and Agriculture Organization reported in 2006 that livestock (principally cattle, pigs, and sheep) account for 18% of the global human-caused CO_2-equivalent emissions (mostly as methane), which is startlingly more than the entire global transportation industry at 14% (per Stern).

STEVEN E. SONDERGARD

More Detail:

CO_2 and CO_2-eq Levels in the Atmosphere

There is currently an estimated 750–800 billion tons of CO_2 in our Earth's atmosphere. Mankind is thought to have increased the concentration from about 280 ppm to about 380 ppm. Therefore, man is being blamed for about 30% of that amount (about 200–210 billion tons in the atmosphere). The amount of annual anthropogenic emissions is not known with great certainty, and various sources estimate different quantities. The USDOE/NETL estimates that 24 billion tons/year of CO_2 were emitted globally in 2004 (this is the latest year for which global information was available). The USDOE/EIA estimates that 28 billion tons/year of CO_2 were added globally in 2004, with 29 billion tons/year estimated in 2005. The IPCC estimates that 30 billion tons/year of CO_2 were emitted globally in 2004. The variability from these sources reflects the uncertainty of the data. (Note that a CO_2 emissions quantity of 30 billion tons per year equates to 8.2 billion tons per year of carbon and to approximately 31 to 35 billion tons per year of CO_2-eq emissions.)

It is reported that the U.S. emits just over 6 billion tons of CO_2 per year. Of the global CO_2 emissions, over 40% are attributed to the U.S. and China, with China believed to have just surpassed (2007) the U.S. because of the rate they are constructing new coal-fired power plants. Therefore China is now emitting over 6 billion tons of CO_2 emissions per year. Indonesia is also rising very fast in their impact on global CO_2 emissions, and they appear to be approaching third place mostly due to the impacts of their significant deforestation activity.

It is estimated that nature generates from thirty to fifty times as much greenhouse gases as does man. However, most of this is in the form of water vapor. The largest contributor of greenhouse gases is due to the heating of the oceans by the Sun, which causes water vapor to emerge and form clouds. But the water vapor content in the atmosphere varies significantly and often. This variation is frequently due to the saturation of atmospheric humidity, resulting in rain and snow. Water vapor also has a lower density than does our nitrogen + oxygen atmosphere. Therefore, the pockets of higher atmospheric moisture have an impact on the weather by causing lower pressure zones and wind. Clouds also impact the weather and the greenhouse effect. Clouds have the dual ability for both warming and cooling. High-altitude thin clouds warm the atmosphere during the daytime and also the nighttime, while lower-altitude, dense clouds have the impact of cooling the atmosphere during the daytime (by reflecting considerable sunlight back into space) and warming the Earth during the nighttime (by absorbing the infrared radiation from the surface). The longer-term impact on climate change of water vapor in the atmosphere is extremely difficult to determine. It is also difficult to ascertain how water vapor acts as a long-term feedback mechanism.

Global anthropogenic CO_2-equivalent emissions in year 2000 were 42 giga tonnes (billion metric tons) per the Stern Report. Of this quantity, the industrial sector contributed 14%, the power sector contributed 24%, the transportation sector contributed 14%, buildings contributed 8%, other energy sources contributed 5%, agriculture contributed 14%, land use contributed 18%, and waste contributed 3%. This means that the combined global industry (including the power sector) contributed 38% of total man-made global GHG emissions. Individuals reportedly contributed about 39% of man-made global emissions, which is slightly more than industry, through transportation vehicles, building comfort, waste generation, and food consumption—this being traced back to the agricultural emissions required for producing the food. The remaining 23% is due to other energy sources and deforestation activities (for homes, business, and agriculture). The startling discovery (for most people) is that individuals have a direct influence on roughly half of the overall greenhouse gas emissions, and that implies that individuals would need to share roughly half of the overall pain for turning things around.

STEVEN E. SONDERGARD

For the U.S. for 2005, the EPA stated that 8.0000 billion tons of human-caused CO_2-equivalent (by one-hundred-year GWP) greenhouse gas emissions were emitted. (The accuracy indicated here is what was actually reported by the EPA.) U.S. emissions fell another 1.1% between 2005 and 2006 according to the EPA. The breakdown is that residential accounted for 17%, agricultural accounted for 9%, transportation accounted for 28%, commercial buildings accounted for 17%, and industry accounted for 28%. It is not common knowledge that the greenhouse gas emissions contributed by U.S. industry (2.25 billion tons in 2005) have been on a steady decline for more than a decade. An estimated 1.5 billion tons of CO_2 came from U.S. coal-fired power plants. It is also worth highlighting that industry within the U.S. contributes less than 30% of total human-caused U.S. GHG emissions. Specifically excluded by the EPA were the consideration of water vapor, human and animal respiration, military activity, and forest fires (human-caused or otherwise). The bulk of this information was obtained from the EPA Web site.[1]

The greenhouse gases that man emits are believed to have a much longer-term impact in the atmosphere than most natural greenhouse gases. This is the basis for determining the one-hundred-year GWP of greenhouse gases. The IPCC has stated that it is "very likely" that human activities are driving global warming. The "human activities" include many activities that may not necessarily be considered bad; at least they would usually be considered to have *some* degree of positive value. In some semblance of decreasing value, these aspects include human respiration, emissions from livestock raised for food, deforestation for farming and habitation, electricity generation, heating/cooling of buildings, transportation vehicles, and stationary industrial sources that provide jobs and improve the quality of life. Reducing many of these aspects, even if proven to be required, would be difficult for most people to accept.

There is strong evidence that there are natural reduction mechanisms that work to deplete the amount of greenhouse gases in the atmosphere. Without some kind of natural reduction mechanism(s), the estimates of anthropogenic emissions would double the atmospheric CO_2 concentration in twenty-seven to thirty-three years. The largest natural reduction mechanism is the oceans, with land masses being the second largest reduction mechanism. Both oceans and land combined are believed to currently soak up roughly half of all anthropogenic emissions. It is also believed that the ability of the oceans and land to soak up CO_2 emissions is diminishing, by some currently unknown degree. This would effectively increase the overall feedback mechanism of anthropogenic CO_2 emissions. Alternatively, other mechanisms appear to be at work to reduce atmospheric greenhouse gases. These other mechanisms might further sequester CO_2 emissions by something like 20% to 30%. After the combined effect of these reduction mechanisms, approximately one-quarter to one-third of the current anthropogenic GHG emissions are then left to accumulate in the atmosphere.

The Observed Reduction in U.S. Industrial Emissions

The GHG emissions from U.S. industry have been on a steady decline for over ten years. The USEPA acknowledged this trend in their GHG National Inventory Report for 2005. In the 2006 EPA report, U.S. industrial CO_2 emissions fell slightly, but overall GHG emissions increased slightly. There are several driving forces behind this trend of declining industrial emissions. One of the largest is energy efficiency improvements. Many energy efficiency measures have been pursued by industry. Benchmark studies of energy usage over the last couple of decades indicate that industry on average has had a steady improvement in energy efficiency that has resulted in an overall improvement of around 20% over the last couple of decades. This directly translates into a reduction in emissions relative to industrial throughput.

One example of energy efficiency is the improvement of industrial fired-burner technology. Replacing burners has been one of the easiest and cheapest ways to reduce NO_x emissions. The average flame temperature does not really change in new burners. But the reduction of NO_x emissions results from a reduction in the *peak* flame temperature from a burner. An added benefit is that new burners are usually more thermally efficient because they control air leakage better and there is less maintenance required. With rising fuel prices it has been economical for industry to change out existing burners in both fired-boilers and fired-heaters. Emission regulations are also requiring many flue gas streams to be monitored for various components. This monitoring has then been extended by industry to include monitoring for excess oxygen in flue gas streams, which can additionally be used for controlling excess combustion air, resulting in improved fuel efficiency as well. Changing out burners and monitoring flue gas streams has been widely pursued by industry in North America. The result has been both beneficial for the environment as well as economic for businesses.

Another large driving force of reduced U.S. industrial CO_2 emissions is that North American industry is moving manufacturing plants overseas. This is driven by both economics and more lenient emission regulations in developing countries. Regrettably for worldwide emissions, most countries where manufacturing plants are moving to, such as China and India, do not have emissions regulations that are as strict as the U.S. or Europe, and worldwide emissions are not usually reduced as a result.

Unfortunately, there is another negative effect that multiplies the global emissions when manufacturing relocates overseas. The power plants that supply power to the relocated industry also have less stringent emissions regulations. The power plants predominantly generate electricity from coal, particularly in China and India, and they have increased emissions relative to power plants in North America and Europe. This is an example of how worldwide emissions can actually increase, even significantly, due to the enactment of stringent regional emission regulations that are inconsistent around the globe.

Rising Ocean Temperatures
Increase Greenhouse Gases

Increased atmospheric CO_2 concentrations are at least partially, and probably substantially, caused from increased ocean temperatures due to an increase in both the total solar effect and warmer atmospheric temperatures. The increased ocean temperatures would drive more dissolved CO_2 from the ocean into the atmosphere.

An increase in atmospheric water vapor would also be expected from increased ocean temperatures. The water vapor in the atmosphere increases by about 4% for every 1°F rise in the average ocean temperature, per the Clausius-Clapeyron relation (the relationship between the temperature of a liquid and its vapor pressure). Therefore, it is estimated that the water vapor above the oceans during the twentieth century could have increased about 5%, based on calculated and measured changes in ocean temperatures.[2] Note that this increase alone might equate to an increased water vapor content over the oceans that could have a greenhouse impact as large as the entire atmospheric concentration of CO_2. This impact would necessarily reduce the direct-only impact of CO_2 on temperature. Countering the increased atmospheric water content, recall that increased low-elevation clouds would reflect more sunlight and keep the temperature from rising quite as much.

Volcanic eruptions

Plate tectonics has a strong influence on natural volcanic eruptions. These volcanic eruptions combined with the rate of weak carbonic acid rainfall and rock weathering impact the Earth's CO_2–silicate/carbonate cycle. This cycle is very long, at maybe 100 million years. In an average year, it has been calculated that volcanoes emit roughly 0.3 to 1.0 billion tons of CO_2. This quantity is not large relative to other emitters. However, relatively sudden impacts from volcanic eruptions do occur that are much higher than this average. These eruptions can

STEVEN E. SONDERGARD

suddenly add multiple atmospheric pollutants, CO_2, sulfur dioxide, and particulates and thereby impact the climate. Some of the gases from volcanoes can act to warm the atmosphere. But overriding the warming impact are the aerosols and particulates from volcanic eruptions that can have a major cooling impact. For example, the eruption in 1815 of Mount Tambora spewed such a large amount of aerosols and particulates into the atmosphere that it caused the following summer of 1816 to be known as "the year without a summer" around the globe. The eruption in 1883 of Krakatoa has been estimated to have caused a drop in the global temperature of about 1°F for a year, and the eruption in 1991 of Pinatubo is reported to have caused a drop in the global temperature of almost 1°F for two years.

The Archaea Domain

The greenhouse gas contribution by the Archaea domain does not appear to be widely recognized. Domains are categories of life such as Bacteria and Eucarya, which includes plants, animals, protists, and fungi. The Archaea is a domain that is divided into Crenarchaeota, composed of heat loving forms, and Euryarchaeota, mainly forms that produce methane as a biological byproduct. Archaea were only recently discovered in the 1980s, and it was one of the most important discoveries ever made in the understanding of the environments that can support life.

Most Archaea are anaerobic (they can only live in the absence of oxygen). Most Archaea live in porous rock far underground. These organisms live totally independently of solar energy. Archaea are termed autotrophs, which means they can produce organic material from inorganic compounds. They produce methane from the hydrogen and carbon that are dissolved or present in the rock, and they could possibly work in concert with the biogenic activity on coal, shale, and kerogen deposits. It is this domain of living organisms that is attributed to creating much of the natural gas beneath the

Earth's surface. Microbes of this type are believed to be the primary producer of coal bed methane (CBM) and possibly the unconventional natural gas found within shale deposits.

The total biomass of all microorganisms underground is currently believed to be several times more than all the combined living matter on the surface. Archaea therefore produce a lot of methane—naturally. When underground methane has an impermeable cap rock over it, the methane will collect under the cap rock over time, and man can recover it in quantity. Otherwise the methane just gradually percolates up through porous rock and eventually into the atmosphere—naturally. In drilling for underground natural gas, mankind simply allows the methane trapped in underground pockets to reach the surface faster than nature would otherwise allow.

Methane currently has a concentration in the atmosphere of 1.77 ppm. This is approaching three times the pre-industrial level of about 0.72 ppm. With the one-hundred-year GWP for methane at twenty-three, the pre-industrial effect of methane alone was about 16 ppm of CO_2-eq versus today's effect at 40 ppm of CO_2-eq, without adjusting for the GHG interference effects. It is also worth noting that the atmospheric methane concentration stopped increasing in about 1990. Needless to say, understanding the impact of methane is important to understanding our future overall CO_2-eq concentrations.

The Max Planck Institute Indicated That Living Plants Emit Methane

In January 2006, *Nature* reported the findings of Frank Keppler (an environmental engineer at the Max Planck Institute for Nuclear Physics in Heidelberg, Germany) that living plants contribute 10% to 30% of the global methane entering the atmosphere (projected to be between 63 and 236 million metric tons per year, with plant debris adding another 1 to 7 million metric tons per year).[3] Keppler and his team measured the amount of methane given off by plant debris in methane-free chambers. They bombarded the plants with gamma radiation to sterilize bacteria and rule out bacterial impacts. But the team did not see much variation in methane produced from sterilized and unsterilized plants. A gram of dried plant material generated 3 nanograms of methane per hour at 30°C. Living plant material released as much as 370 nanograms of methane per gram of plant material per hour. Methane emissions tripled when plants were exposed to sunlight. At the time the *Nature* article was written, they did not yet know the mechanism that produced the methane but speculated that pectin in the plant cell walls plays a part. The tests were repeated to confirm the startling quantities of methane being produced. The tests were performed on several types of plants including maize, basil, and wheat. The effect of this finding was previously completely unknown, and it was not recognized in any climate models or other considerations prior to this reporting.

This finding is surprising and even controversial, and it is viewed by the mainstream scientific community as a probable mistake. The main argument against it is that it doesn't correlate with what we already know about chemistry. Methane is thought to be only formed in anaerobic conditions. Other researchers have been constructing their own experiments to confirm or deny the Keppler findings. At the moment it doesn't appear likely, but if this finding holds up, it would lead us to question other methane sources in the methane balance; plus it would have implications on climate policy that would promote the use of plants (forestation) as CO_2 offsets and sinks. If this finding holds up, it may also help explain a cause of the methane concentration variability found in ice cores. The Keppler team finding would also help explain another finding reported in *New Scientist*, 2005, that satellites have observed large methane plumes over tropical rainforests, which indicates tropical rainforests may be emitting more methane than previously thought.[4] If linked to deforestation, the finding might also help explain why global methane levels have recently stopped rising.

A more recent study reported online by *New Phytologist*, 2007, contends the Keppler team's findings are not correct.[5] Biologists grew maize, basil, and wheat using heavier carbon-13 instead of carbon-12. This isotopic labeling allowed a determination of where any resulting methane came from. The biologists who performed these recent tests found no significant emissions of methane from the plants they grew, even though they looked at a large amount of plants. They also faulted the methods used by the Max Planck chemists as reported in *Nature*. Now, the two research groups plan to collaborate and re-run experiments to see why they had such widely different results, and they hope to ultimately determine whether living plants emit methane or not.

GHG Emissions Directly from Human Population

Interestingly, global human respiration (breathing) emits between 5% and 10% of human-caused CO_2. This is rarely, if ever, considered in emission calculations—which is a little surprising as it is rather significant. Those who ignore it may simply want to avoid the controversy and the perceived insensitivity that raising it might cause. Nonetheless, it is still a contributing factor to human-caused CO_2 emissions, and understanding its significance allows us to better understand the magnitude of the overall GHG issue and allows us to better develop adequate overall policy. The following calculation is an indicator of the direct impact of human population on global greenhouse gas emissions.

The Quantification of Human Respiration

- Assume there are 6.5 billion people on Earth (the approximate population in 2005).
- Some are old, some are young, some are exerting, and some are sleeping.
- Assume the average respiration rate is fourteen breaths per minute.

STEVEN E. SONDERGARD

- There are approximately 525,600 minutes in a year (365*24*60).
- There is about 21% oxygen in the air.
- Assume that 25% of the oxygen is replaced with CO_2 by each breath.
- Assume the volume of air that is exhaled averages 0.4 liters per breath.
- Assume the exhaled breath averages about 95°F.
- There are 28.31 liters in a cubic foot.
- There is 379 cu.ft/#mole of a gas at one atmosphere pressure and 60°F.
- CO_2 has a molecular weight of 44.
- There are 2,000 pounds per ton.

6.5 billion * 14 * 525,600 * 21% * 25% * 0.4 / 28.31 / 379 *44 / 2000 * (60+459) / (95+459) = **1.93 giga tons/year**

With an estimated 24 to 30 giga tons/year of CO_2 emissions worldwide in 2005, the amount estimated to be contributed by human respiration alone is 6.4% to 8.0%.

Indirect GHG contributions also occur from the respiration of human pets. This may be an additional CO_2 contributor of maybe one-fourth to one-third of the human respiration. Human and pet respiration combined might therefore be around 10%. Further, the number of animals raised for food and the number of animals raised for clothing must be considered as well. Together, human respiration plus human-caused animal respiration appears to account for 15% to 20% of all CO_2 emissions.

Several farm animals also emit significant quantities of methane, in addition to CO_2. The total GHG emissions resulting from farm animals can be very significant. The United Nations Food and Agriculture Organization reported in 2006 that livestock (principally cattle, pigs, and sheep) account for 18% of the global anthropogenic CO_2-equivalent emissions (mostly as methane), which is startlingly more than the entire global transportation industry at 14% (per the Stern Report).

Apparently over one-quarter of global CO_2-eq emissions are therefore the direct or indirect result of human-caused animal bodily functions! These emissions will not be reduced by a cap-and-trade mechanism. On the contrary, these emissions would actually increase linearly with any human population growth, assuming the growing population would have similar eating habits. The reverse would also be true. A reduction in global population would have a beneficial impact on global greenhouse gas emissions. A reduction in population would have the added benefit of having less strain on our limited food and water resources.

Another way to look at it is that without a population reduction, the highest possible global GHG emissions reduction would be less than 75%—and that would only be possible with *absolutely zero* global emissions from industry, electricity generation, heating and cooling of buildings, cooking, lighting, and transportation. This is an important point. Population increases could only make it worse. It is unrealistic to expect zero emissions from these categories, and pursuing emissions reductions that would even approach a 75% reduction would be expected to initiate a major societal collapse and/or a severe reduction in population all by itself. Note: the complete obliteration of GHG emissions from the global power and industrial sectors would only get halfway toward a 75% reduction target.

8. WHAT ABOUT TEMPERATURE PROJECTIONS, GHG THRESHOLDS, AND EXPECTED OUTCOMES?

Focus: The IPCC temperature projections are relayed. The most prescribed thresholds and forecasted outcomes from increasing temperatures are evaluated. Rising sea levels and historical mass extinctions of species are discussed.

Layman's Brief:

The dominant share of information on the magnitude of the problem describes the potential outcomes from Global Warming as calamitous. The forecasted outcomes include droughts, reduced water availability, reduced food production, famines, weather pattern changes, increased hurricane activity, ocean current changes, and coastal flooding from ice cap melting. A threshold tipping point for initiating many of these outcomes is often stated to occur when the average global temperature increases by a couple of degrees Centigrade. This temperature increase has then been translated into CO_2 or CO_2-eq target levels that are described as global maximum targets.

There are others who convey that these disastrous outcomes are overstated and that these forecasts appear to be chiefly alarmist in nature and simply promote action toward a specific agenda. Indeed, there is little argument that the Earth has been through many global warming cycles that are at least as severe as the present warming trend. It must also be acknowledged that several known natural causes of global warming do exist, including solar variation. As outlined, overall feedback mechanisms may be overstated, particularly when considering solar variability. The shape and magnitude of commonly projected temperature trends can therefore be described as overstated. The threshold CO_2 levels required to maintain less than a two-degree-Centigrade rise also appear to be overly restrictive. Opponents contend the calamitous outcomes described are often exaggerated from the described cause, and many of these outcomes can be expected to occur naturally. They declare that anthropogenic Global Warming doesn't appear to be nearly as bad as it is often conveyed. A balanced, objective examination of the data should help reveal a realistic answer.

Quick Reference:

- From a historical perspective, global warming trends are a common part of cycles that have occurred many times in the past, and these past cycles are not abnormal. In addition, the current observed warming appears to be within past warming cycle magnitudes. However, this does not mean the current warming trend is only the result of natural causes.
- The IPCC and other reports project an increasing temperature trend of exponential nature. For this forecast to become reality, the atmospheric CO_2 concentrations must not only increase exponentially, but overall feedback

mechanisms must also be positive, and their impact must increase exponentially as well.

- The IPCC, the Stern Report, and others have relayed there is a threshold at 450 ppm to 550 ppm of CO_2-eq where a tipping point exists and where multiple climate calamities could start to occur. These thresholds are quoted frequently, and it is often stated that there is a strong scientific basis for them, yet a strong scientific basis for these hard thresholds appears to be virtually absent. The actual threshold is much fuzzier than these 450 ppm or 550 ppm threshold targets would lead one to believe. The 550 ppm target is roughly double the pre-industrial level of CO_2. It appears this was set as a target to initiate action. The 450 ppm target is forwarded as a number desired to keep temperatures from rising more than 2°C (3.6°F) from pre-industrial levels. Two comments can be made about these threshold levels. First, the most realistic concentration threshold appears to be higher than 450 ppm or even 550 ppm of CO_2-eq to remain under the 2°C global temperature increase from human causes. Second, many of the calamities forecasted to accompany this 2°C temperature rise are quite pessimistic, and others will occur whether the Earth remains under these concentration thresholds or not.

- There are many forecasts of global climate calamities that incite emotion or instill fear. These forecasts include droughts, reduced water availability, reduced food production, famines, weather pattern changes, increased hurricane activity, ocean current changes, and coastal flooding from ice cap melting. But others believe these forecasts lean toward being alarmist in nature, and they offer some seemingly valid reasons for this belief.

- The IPCC AR4, 2007, report states that the ocean levels rose by a total of 6.7 inches in the twentieth century. Since the Arctic ice sheet is floating, if it melted, it would not add to the ocean level. However, if the Antarctic or Greenland ice sheets were to melt, they *would* increase the ocean level. The IPCC report also notes that if the Greenland ice sheet were to completely melt, requiring melting for multiple centuries to millennia, the oceans would rise up to seven meters (up to twenty-three feet). It is estimated that if *all* of Earth's polar ice were to melt, it would raise the ocean level about 200 feet; a range of calculations lie between 186 feet and 275 feet, relative to today. It should be understood that there is strong evidence indicating ocean levels naturally rise and fall over 410 feet to as much as 470 feet cyclically between ice ages. Past evidence also suggests the oceans will probably rise at least another ten feet and likely seventy feet (or maybe more) in the current interglacial—naturally.

- The polar bear has often been cited as a poster child for being under increased stress due to global warming. But this bear has adapted and survived past ice age cycling, or it would not be alive today. In fact, the polar bear is reported to be currently thriving nicely and increasing in number. All other life forms reported as being threatened by global warming have also been through many ice age cycles and ocean level changes previously as well, or these species would not have survived until today either.

- The risks of forecasted calamities are very much unknown, yet they definitely pull at our emotional heartstrings. The acceptance of these risks, large or small, is likely influenced by one's risk tolerance and one's perceived personal, family, or country impact. But how much should we allow our personal fears to influence our longer-term and over-arching thinking? It is suggested that intellectual maturity should overrule our emotions.

- Several prehistoric extinctions of species have been correlated to peak levels of CO_2 in our atmosphere. These past extinctions correlate to CO_2 levels of at least 900 ppm and usually 2,000 ppm to 4,000 ppm. In addition, the timing of several of the past mass extinctions correlates with the age of large basalt formations around the globe, and therefore major volcanic activity. Plus, twelve of fourteen studied mass extinctions correlate with data suggesting there were poorly oxygenated oceans at the time of the extinctions. Other evidence suggests the cause and effect of these extinctions may be due to an outside influence.

More Detail:

IPCC Temperature Projections

The IPCC temperature projections were introduced earlier in Chapter 6, but they will now be discussed further. Remember, these temperature projections are based on AOGCM models. As early as the 1990 IPCC report and the 1995 IPCC conference in Rome, the IPCC concluded there were clear signs of man-made climate changes and that a cause of it was pollution from cars, trucks, factories, and power plants.

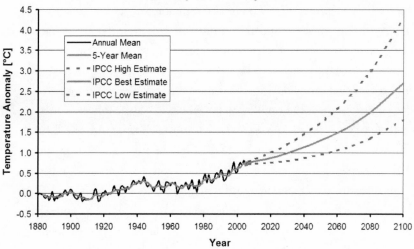

Figure 8.1. IPCC projections of temperature increases due to global warming with past temperature anomalies by NASA GISS (Goddard Institute for Space Studies). Note the shape of the curve is exponential in nature, which is in contrast with the logarithmic curves presented earlier. (Adapted from: IPCC, AR4, 2007 temperature projections)

Figure 8.1 above shows the historical temperature anomalies along with the future IPCC projections. You can read much more about these facts and statistics from either the IPCC Web site itself or the National Climatic Data Center (NCDC), which is the world's largest active archive of weather data. These temperature projections are forecasted to be accompanied by rather scary outcomes, including increasing tropical storms and tornadoes, killer heat waves, above-average rainfalls, rising ocean levels, increasing tropical diseases, and even species extinctions. Scientists at the National Oceanic and Atmospheric Administration (NOAA) report that these weather extremes are consistent with what they expect global climate change is likely to bring.

It can be easily observed that the shape of these curves projected by the IPCC for temperature are dramatically different than the shape of the logarithmic saturation curves for temperature presented earlier. For the global average temperature curve to bend upward as the IPCC projects, there are three major factors which must be true and that must occur. First, it must be true that the atmospheric CO_2 concentration is the dominant driver of global warming. This assumption can, at the very least, be reasonably questioned, as highlighted earlier, considering the impact of both water vapor, solar variability, and orbital oscillations. Second, the atmospheric concentration of CO_2 must increase *exponentially* with time into the future. This assumes the demand for fossil fuels will continue to produce this exponential increase in CO_2, that conservation and mitigation measures will have little influence on CO_2 emissions, and that crude oil production will be able to keep up with this exponential fossil fuel demand. These assumptions are covered in more detail in later chapters, but they can each be seriously questioned. Third, the overall impact of all combined feedback mechanisms of global warming must not only amplify the global warming further, but it must do so with *exponential* impact. Although feedback mechanisms could very well be changing and increasing in their positive feed-

STEVEN E. SONDERGARD

back impact, their sensitivity would have to be very high in order to yield an exponential effect. This exponentially-increasing sensitivity does not appear to be supported by known data. Either one of these two exponential requirements alone would only straighten the logarithmic saturation curve and result in a roughly linear temperature increase versus time. However, exponential CO_2 concentration increases and exponentially increasing feedbacks must *both* be realized for the curve to overcome the logarithmic effect and bend upward as the IPCC has projected. It is indeed possible for this to be the case, but it does not seem very probable.

The magnitude of the IPCC-forecasted temperature rise caused by GHGs can also be questioned, in large part from the questionable shape of their curve. This was mentioned earlier in Chapter 6 as well. What goes unstated in most reports is that a global warming might be expected to occur even without anthropogenic causes and several of the scary outcomes conveyed can be expected to occur naturally anyway. This aspect will be covered in more detail following the discussion of 450 ppm and 550 ppm thresholds.

The 450 ppm and 550 ppm Thresholds

The IPCC and others have relayed there is a tipping point at either 450 ppm or 550 ppm of atmospheric CO_2-eq levels where multiple climate calamities would result in negative outcomes for humanity. These threshold numbers are actually quoted often as having a strong scientific basis, and the threshold numbers are rarely questioned in the literature. The IPCC lists 550 ppm of CO_2-eq as the threshold. The Stern Report lists 450 ppm CO_2-eq to 550 ppm of CO_2-eq as a threshold. The Union of Concerned Scientists lists 450 ppm of CO_2-eq as a threshold. Unfortunately, a strong scientific basis for these thresholds appears to be virtually absent.

These thresholds will now be examined a little more closely. The 550 ppm target was set to initiate action. It is a number set at roughly double the pre-industrial level of CO_2. The 550 ppm number can also be roughly obtained by starting with the CO_2 level from a couple of decades ago and projecting the present rate of increase of 2 ppm per year to the end of the twenty-first century. The 550 ppm of atmospheric CO_2-eq roughly equates to about 450 ppm of CO_2 concentration. The 450 ppm target is a number projected by some to keep temperatures from rising no more than 2°C (3.6°F) above pre-industrial levels. Actually, this temperature increase is much fuzzier than the target would lead you to believe. In addition, many of the calamities forecasted to accompany this 2°C temperature rise are quite pessimistic, and some of these forecasts will occur whether the Earth remains under the concentration threshold or not.

Many reports on climate change reference one of these 450 ppm or 550 ppm of CO_2 or CO_2-eq thresholds and further indicate that levels beyond their specifically stated threshold must be avoided at all costs—even if the cost is civilization itself.

These threshold numbers of 450 ppm and 550 ppm can be examined further and logically broken down into two parts. The first part correlates the atmospheric CO_2 level to a global temperature increase. The second part of the breakdown relates the projected temperature increases to specific global outcomes, and it will be covered shortly.

For the first part then, the correlation between 450 ppm or 550 ppm of atmospheric CO_2-eq levels and a 2°C global temperature rise can be reconstructed by starting with either the most-probable direct logarithmic effect or the derivative of the Stefan-Boltzmann law. These techniques were described earlier. The logarithmic effect suggested that a doubling of the CO_2 concentration might have the direct impact of increasing the temperature by 1° to 1.5°F. The derivative of the Stefan-Boltzmann law may suggest a doubling of the CO_2 concentration will have the *direct* impact of increasing the temperature by not more than 2°F. The more aggressive 2°F is what will be initially used here. Additionally, an overall feedback multiplier of 2.0 is initially assumed. The observed temperature impact would also be impacted by the effect

STEVEN E. SONDERGARD

of greenhouse gas interferences. A relatively low interference factor of 10% will be initially used resulting in a multiplier for the reduced greenhouse effect of 90% due to greenhouse gas interferences. The calculation for an observed temperature increase with a doubling of CO_2 concentration thus yields an answer of 2.0°C (this can be calculated by: 2.0°F * 2.0 feedback * 90% = 3.6°F, which is 2.0°C). This answer correlates with the lower end of the range stated in the IPCC AR4, 2007, where a temperature increase due to a doubling of GHG levels is forecasted to be 5.7°F (with a range of 3.7°F to 7.9°F). A corresponding correlation between the CO_2 concentration threshold and an observed temperature increase can also be developed. The concept of climate sensitivity alone suggests the correlation is logarithmic. Such a correlation would be: 2°F * 5/9 * 2.0 feedback * 90% = K_1 * ln (560 ppm / 280 ppm). The constant K_1 becomes 2.885 in this example. A 560 ppm threshold would allow an *increase* in the atmospheric CO_2-eq concentration of 280 ppm relative to the accepted pre-industrial levels of 280 ppm, in order to keep the temperature from rising more than 2°C.

Alternative assumptions can also be made to observe their impact. It is known that the interferences of greenhouse gases can reduce the greenhouse effect by more than the 10% initially assumed. What would be the impact of using an average interference of 25%? In addition, the overall feedback multiplier for anthropogenic causes could be closer to 1.7, as outlined earlier. The expected temperature rise for a targeted level of 560 ppm CO_2-eq could therefore be 1.42°C, or 2.55°F (this can be calculated by: 2°F * 5/9 * 1.7 feedback * 75% = 1.42°C). Then, using a similar logarithmic correlation again to determine a threshold for a 2°C temperature rise gives a **743 ppm** CO_2 threshold (this is calculated from: 1.42°C = K_2 * ln (560 ppm / 280 ppm). The constant K_2 becomes 2.049 here. Then solving for x_2 in the equation: 2.0°C = 2.049 * ln (x_2 / 280 ppm) yields 743 ppm for x_2.

The higher threshold determined from these alternative assumptions would allow an *increase* in atmospheric GHG levels of 463 ppm relative to accepted pre-industrial levels (calculated by: 743–280) before the 2°C increase would result. The difference between the two calculations is 60% to 70%, and it indicates the CO_2 threshold for a 2°C temperature rise could be significantly higher than what is promoted in prominent literature.

Alternatively, if the stipulated logarithmic relationship is instead linear due to exponentially increasing feedbacks, a 2°C temperature rise correlates to a **675 ppm** CO_2 threshold (this is calculated from: 2°F * 5/9 * 1.7 feedback * 75% = K_3 * (560 ppm - 280 ppm). The constant K_3 becomes 0.005 here. Then solving for x_3 in the equation 2°C = 0.005 * (x_3 - 280 ppm) yields 675 ppm for x_3. If the *direct* temperature impact for a doubling of GHGs approaches 1.5°F, or even 1°F as the possibility was presented earlier, the CO_2 threshold for a 2°C temperature rise could even be well over 1,000 ppm.

In reality, none of the above calculations should be considered correct. What these rough calculations should reflect is the fact that there is still considerable uncertainty in any CO_2-eq concentration threshold and a threshold of 450 ppm or even 550 ppm appears to be too low. A threshold near 550 ppm is only possible if CO_2 is the only cause of warming, the relationship is linear, and there are no GHG interferences. It should actually be heartening to the reader that the GHG threshold might be higher, because this could become quite important for civilization, particularly if mankind cannot easily curb anthropogenic GHG emissions. A later chapter will deal with the *probability* of curbing anthropogenic GHG emissions.

Forecasted Outcomes from Increasing Temperatures

What are the outcomes projected to occur when specific temperature increases are reached? There are several literature sources that have stipulated the impacts that might occur for various temperature increases of one to a few degrees. The Stern Report is among the most widely referenced of these sources, and it highlights the majority of these outcomes. The climate calamities associated with the thresholds include species extinctions, droughts, reduced water availability, reduced food production, famines, weather pattern changes, increased hurricane activity, ocean current changes, and coastal flooding from ice cap melting. Each of these effects will be discussed in turn. Opposing comments about each outcome that might counter commonly-held beliefs are also included. Even among those who forecast similar calamities, there are inconsistencies in the severity of outcomes to global temperature increases. This variability alone supports that there is still a lack of certainty in these forecasted outcomes.

An additional assumption is apparent by many forecasters that there is a single cause-and-effect relationship between atmospheric greenhouse gas levels and climate calamities and that humans are the only cause that could result in these climate calamities. However, it would be naïve to assume that our climate will not change from natural causes. The Earth's temperature could easily change by 2°C naturally. Not only that—it should be expected.

One of the fallacies of many who project calamities is that the question is seemingly never asked about what calamities can be expected to occur even if mankind *does* reduce GHG emissions as prescribed, with a resulting stabilization of the atmospheric CO_2 level occurring at or below the targeted maximums. The answer is that some of these forecasted outcomes are likely to occur naturally whether the world meets the prescribed anthropogenic GHG targets or not. The historical climate variability and the rapid changes in our past climate (both observed in the ice core record) lead to the belief that our climate should be expected to change naturally and many forecasted calamities will occur anyway.

The Stern Report specifically forecasts that 15%–40% of species extinctions would be initiated with only a 2°C average temperature increase. Yet, this forecast does not appear to align with historical evidence or even common sense. For one, many greater temperature changes and increases have occurred in the past, including diurnal variability, seasonal variability, and the Holocene Maximum, without observing these species extinctions. There is not even any evidence to suggest that the much greater glacial-to-interglacial temperature swings have resulted in extinctions such as the Stern Report forecasts. Further, the impact of the 2°C average global temperature increase on most species is diminished when considering most of this temperature increase will occur near the poles and at night. The average temperatures near the equator and during the heat of the day are not expected to change nearly as much.

Associated with species extinctions is the projection of extensive and irreversible damage to coral reef ecosystems after only 1°C of temperature rise. Opponents confess that coral reefs may indeed become stressed, but they also argue that this outcome can be expected to occur naturally, even without anthropogenic emissions. The argument by opponents continues in stating the climate cannot be expected to remain constant. Ocean levels have naturally risen and fallen hundreds of feet with every interglacial period, yet coral have survived through these relatively severe climate changes of the past. Future climate cycles are not expected to be much different than these past natural cycles. Sensitive coral can be expected to either adapt or initiate existence at a nearby home where the water depth and water temperature would be just right for them—as they have done through multiple past glacial-to-interglacial cycles. The risk of coral extinctions is therefore considered to be much lower than what is implied in the forecast.

Another species often cited as being at risk is the polar bear. The polar bear may indeed be forced to adapt, but it is apparently doing just that today. It is reported that the polar bear is currently thriving nicely and its numbers are actually increasing globally. Of the twenty or so distinct subpopulations of polar bears, there is only one or two that are declining—they are near Baffin Bay. The majority of polar bear subpopulations are stable, and two are increasing near the Beaufort Sea.[1] Professor J. Scott Armstrong, a specialist in forecasting at the Wharton School, says, "To list a species that is currently in good health as an endangered species requires valid forecasts that its population would decline to levels that threaten its viability. In fact, the polar bear populations have been increasing rapidly in recent decades due to hunting restrictions. Assuming these restrictions remain, the most appropriate forecast is to assume that the upward trend would continue for a few years, then level off."[2] A similar logic would apply to most other threatened species that applies to the polar bear. Each of these "threatened" species has survived many extreme temperature cycles in the past.

The forecasted outcome of 25%–60% increase in the number of people at risk from hunger with a 2°C rise in temperature appears to be slanted. This increased risk of hunger is purported to occur simply from the current trajectory of increased population, even without a global temperature rise. In fact, the number of people at risk from hunger is expected to increase by about 100% simply as a result of corn-based ethanol production alone, per researchers at the University of Minnesota.

There are projections that water supplies will become short, beginning with a 1°C to 2°C temperature rise. Water shortages are indeed forecasted to occur. But the cause and effect is made as if water shortages would not occur if it were not for Global Warming. Many forward thinkers claim that clean water is projected to become a scarce resource in the future, whether global warming occurs or not. Water tables are being depleted, water runoff should be expected to be variable, population increases will diminish available water, and water is forecast to become scarce, no matter what happens with greenhouse gases.

The forecasts of irreversible melting of the Greenland ice sheet with only 1°C to 2°C temperature rise does not represent a proper accounting of historical ice sheet cycles and their causes. The melting of much of the Greenland ice sheet has occurred with every interglacial cycle on record, and this melting should be expected to occur naturally during the present interglacial as well. As previously discussed, these cycles and the subsequent melting are dominantly caused by natural orbital variations of our Earth, and the evidence suggests this melting is neither abnormal nor irreversible.

The Stern Report makes the statement that temperatures from climate change will take the world outside the range of human experience. This statement appears to be quite alarmist in nature. It could only be expected to become true if it pertains to the distant future, if the population growth continues upward, if crude oil production and other fossil fuel resources last much longer than expected, and if man does virtually nothing to change course or to adapt.

The forecast statement that sea levels will rise, flooding will occur, and people will lose their homes may indeed prove to be accurate. But the stipulated cause and effect may be misleading. Data (detailed in the next section) suggests that sea levels rise and fall naturally and significantly between ice ages, weather patterns naturally change significantly and suddenly, flooding will occur from many causes, and many homes have simply been built in high risk areas. Some argue it surely cannot be expected that all of those who have built in high risk areas should be protected from their own unwitting behavior and even poor decisions. Some people will naturally accept more risk, and there will always be those who push the risk envelope. Still others facing the requirement for change simply refuse to adapt.

A predicted 5% or 10% increase in hurricane wind speed by the Stern Report or the increased hurricane activity reported at Harvard by the IPCC's Kevin Trenberth is not supported by NOAA, who is the recognized world authority on hurricane activity. The conclusion about increased hurricane activity might be supported by a trend of increasing hurricanes that started in about 1970 and that possibly ended in 2005. But climate modeling does not support increased hurricane activity from global warming. NOAA stated in April 2007 that each of their eighteen global climate models consistently predict *reduced* hurricane activity from global warming, due to a robust increase in vertical wind shear in the tropical Atlantic and the eastern Pacific Oceans. Of course, it is statistically possible that a hurricane of high intensity could avoid this increased vertical wind shear, but even when it does occur, it would not necessarily mean the cause is anthropogenic. Further, the corresponding statement that increased hurricane intensity will result in doubling the cost of damages in the U.S. is not a result of a 2°C temperature rise or even increased hurricane intensity, but it can be expected as a result of increased construction in harm's way, plus the escalating costs of construction.

Is it the expectation that mankind should control the natural causes as well as human causes of these calamitous outcomes? If the answer is no, then the magnitude of these forecasts are overstated, and the probability and magnitude of the outcomes from natural causes need to be subtracted from every event category. If the answer is yes, then man would be attempting to take on the role of God, and many people will have a problem with that desire. In either case, it appears that the thrust of describing the forecasted calamitous outcomes is to shock the general public, to justify the significance of anthropogenic Global Warming, to sway mankind to take these matters into his own hands, to circumvent God, and to promote extreme action. Further, the downside(s) of the potential action measures and any overreaction to the problem appears to be completely dismissed by the Stern Report and other forecasters.

Rising Sea Levels from Melting Ice

The IPCC AR4, 2007, report estimates the ocean levels rose 3.0 inches from 1961 to 2003 and ocean levels rose a total of 6.7 inches during the twentieth century. Other calculations indicate a sea level increase of less than one half of an inch in the last century.[3] How much more can we expect our ocean levels to rise? First, the ocean levels are expected to increase about five feet relative to today just from the warming of cold ice-melt water from already melted ice.[4] When this cold water increases in temperature up to the average ocean temperature, an expansion of the water will occur. This expansion will likely take centuries, because the cold water must travel via the ocean current system to warmer climates. The expansion of water in the twenty-first century alone is expected to raise the sea level by about nine inches.

Second, further melting must be considered. Since the Arctic ice sheet is floating, if it melted, it would *not* add to the ocean level. However, if the Greenland or Antarctic ice sheets were to melt, they *would* increase the ocean level. Within the current interglacial period, it is estimated that the world's glacial mass was at its highest level in the late 1700s, near the end of the Little Ice Age. If all of Earth's mountain glaciers were to melt, it would raise the ocean levels by about 1.5 feet.[5] The melting of ice caps and mountain glaciers in the twenty-first century is expected to raise the sea level by about three inches.

It is estimated that if *all* of Earth's current polar ice were to melt, it would raise the ocean level another 200 feet or more, relative to today.[6, 7] Calculations range from 186 feet up to 275 feet.[8, 9] The Antarctic ice sheet would have by far the largest impact on this ocean rise, as it has an *average* thickness of about 7,000 feet.[10] But Antarctica is currently experiencing increased ice accumulation, and this is forecasted to continue such that the resulting sea level is expected to *fall* by about two inches in the twenty-first century from this impact alone.

The IPCC AR4, 2007, report also notes that if the Greenland ice sheet were to completely melt (requiring melting for multiple centuries to millennia), the oceans would rise up to seven meters (up to twenty-three feet). Others calculate the oceans would rise only six meters (only about twenty feet) with complete Greenland ice sheet melting.[11] There is no question that if the Greenland ice sheets were to melt it would indeed be unfortunate for those with homes near the sea level. But a little more perspective is in order. The expected rise in the sea level from the melting of the Greenland ice sheet in the twenty-first century is only about one to one and a half inches. The melting of the entire Greenland ice sheet and the subsequent change in ocean levels is unlikely to occur for many centuries or even millennia—and much of this melting can be fully expected to occur naturally.

The overall combined impact of mountain glacier melting, Greenland ice melting, polar ice cap melting, and Antarctic ice accumulation yields an estimated twenty-first century sea level rise of about twelve inches. Further, our Earth cannot be expected to remain in a relatively constant climate state, and our climate is better viewed as being in a state of flux. Changes in ocean levels are not abnormal. Ocean levels naturally rise and fall significantly between the 100,000-year ice age cycle. Evidence indicates ocean levels have *naturally* risen and fallen as little as 410 feet to as much as 470 feet cyclically between ice ages.[12, 13, 14] This is substantiated from both old shoreline features and from sediment cores below the Greenland and the West Antarctic ice sheets. The ocean levels of 20,000 years ago, at the Last Glacial Maximum, were 400 feet below where they are today.[15, 16] The breakthrough work for this conclusion was made by Richard Fairbanks in 1988, by radiocarbon dating the borings of a species of coral, *Acapora palmata*, off the coast of Barbados. This species of coral is known to live only within a few feet of the ocean surface, so recording the bore depth and the age provides a good understanding of past ocean levels.

Past interglacial sea level data suggests sea levels can be expected to *naturally* rise at least another ten feet to twenty feet, and the increasing trend in the last several interglacial sea levels might suggest that sea levels could even rise another seventy feet or more within the current interglacial period. Humans may simply need to adapt to rising ocean levels and take reasonable protective action measures. With a few feet of sea level rise, realistic action measures could reasonably include levies and dykes, constructing protective sea walls, restricting development in the lowest areas, elevating building foundations, and using better building techniques.

There are several countries and regions that have been cited as being at high risk of rising sea levels, but reasonable action measures could reduce this impact in every case.[17] It has been estimated that the Maldives could lose up to 77% of its land area without action measures, but economic action measures could reduce that to 0.0015%. For Tuvalu, the loss of land could be economically reduced to 0.03%. Vietnam could reduce its land area loss from 15% to 0.02% with economic action measures. It is reported that Micronesia could lose up to 21% of its land area, but economic action measures could reduce the loss of land to 0.04%. Bangladesh could limit its land loss to a miniscule 0.000034% with economic action measures.

A further impact of sea levels rising relative to coastal land masses is that coastal land can actually subside to some degree. This is often the result of water being pumped from below the land mass by humans, and the land settles more as a result. The sea level does not actually rise due to this mechanism, because the land is what sinks, but the effect is the same in that the sea encroaches upon the coastal land area.

Another aspect of melting glaciers is that a lot of glacial mass is currently being removed that has been pushing the land downward for a long time. Once this downward force is removed, the land responds by rising relative to other points on the surface. This is called post-glacial rebound. One result of this is that northern Europe itself is rising somewhat, and this in turn is believed to be causing southern Europe to get somewhat lower, as the tectonic plate tilts back to level itself a little.

For a little wider perspective, geologic information (e.g. ice marks) over the past 500 million years indicates most of this long geologic time span was probably completely free of ice sheets and the oceans had a correspondingly high sea level. This condition probably prevailed until around 3 million years ago.[18]

Clues from Prehistoric Extinctions

Peter Ward, 2007, describes some thought-provoking correlations about global warming that may have significant impacts to our world.[19]

STEVEN E. SONDERGARD

His evidence will be examined next. One finding is that the timing of several of the past mass extinctions correlates with the age of large basalt formations around the globe and therefore major volcanic activity. Additionally, Anthony Hallum and Paul Wignall, 1997, studied fourteen mass extinctions, and twelve of the fourteen correlated with data suggesting our oceans were poorly oxygenated at the time of the extinctions. A main consideration is that several prehistoric extinctions of species have been correlated to peak levels of CO_2 in our atmosphere. But it must be remembered that correlation does not necessarily imply causation. Were these extinctions caused by the increased CO_2 levels, or were the subsequent reductions in CO_2 levels the result of the extinctions and the extinctions were actually caused by something else? Scientists really don't know. However, Peter Ward does propose a possible mechanism for these past extinctions of species.

Ward calls his proposed mechanism the *conveyer disruption hypothesis*. This mechanism from *Under a Green Sky* is briefly summarized here:[20]

> The world has increased volcanic activity, causing large volcanic basalt flood planes that emit CO_2 and methane. This causes the world to warm and the ocean current system to be disrupted. The pole temperatures further warm and approach the equatorial temperatures. This causes a major reduction in wind and ocean surface currents. Bottom water then gets warmer and the reduced mixing reduces the water's oxygen levels. Water volumes decrease with reduced oxygen, and the shallower water become anoxic. The shallower water allows light to get to the bottom water and this combined with the lack of oxygen begins to grow green sulfur bacteria. These bacteria produce toxic amounts of H_2S, at maybe 2,000 times the rate of today. The H_2S is emitted into the atmosphere where it rises to breakdown the ozone layer. The resulting increase in ultraviolet light kills much of the photoplankton. The H_2S in the atmosphere also kills plant and animal life on the surface causing a mass extinction on land.

The following graph (Figure 8.2) was generated by a GCM model called GEOCARB. Note that this graph is not an actual data record. It is the output of a model. Each data point represented by a large circle reflects a past mass extinction. The GEOCARB model was developed by Robert Berner of Yale University. The results of this model are corroborated by examining the observable stomata of well-preserved fossil leaves found in rock strata. Stomata are tiny portals in a leaf that allow CO_2 into the plant and allow water vapor to escape to the atmosphere. When CO_2 levels are higher in the atmosphere, plants do not need as many stomata to survive. When plants do not need the stomata to survive, they produce fewer of them. Counting the number of the stomata in the fossil record therefore provides a relative measurement of prehistoric atmospheric CO_2 levels.

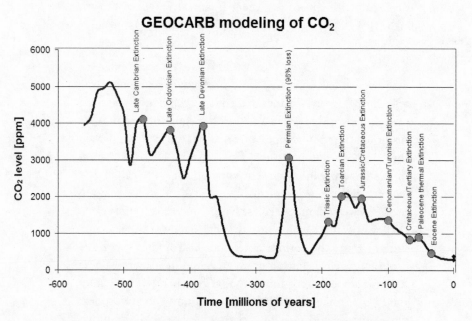

Figure 8.2. A GEOCARB model output of past CO_2 levels over millions of years. Note the correlation to past extinctions. (Adapted from: *Under a Green Sky*, Peter D. Ward, Smithsonian Books, 2007, figure 6.1, p.135)

STEVEN E. SONDERGARD

Several points can be made about this proposed mechanism that might help provide a balanced view. First, it should be noted that the majority of the past 600 million years had a CO_2 concentration that was an order of magnitude higher than it is today. Second, the fossil record suggests that there have been very large, cold-blooded creatures in the past that assuredly needed warmer average temperatures to exist; plus, the dating of these creatures correlates with higher atmospheric CO_2 levels. It seems plausible that higher-concentration atmospheric greenhouse gases could have overpowered the seasonal and Milankovitch cycles enough to reduce or prevent ice formation during these epochs. A realistic conclusion is that higher atmospheric greenhouse gases caused this probable global warming effect. Further, it also seems likely that these higher atmospheric greenhouse gases included higher average concentrations of water vapor. Third, every one of these past mass extinctions were all natural in their cause. Fourth, if these past mass extinctions had not occurred, you and I would probably not be here, and you would not be reading this today! Fifth, the past extinctions of species correlate to CO_2 levels of at least 900 ppm, and the extinctions usually occurred at a CO_2 concentration of between 2,000 and 4,000 ppm. If CO_2 concentration is indeed the cause of the extinctions, these magnitudes could indicate that atmospheric CO_2 thresholds might actually be multiple times higher than what many have stipulated. In addition, there may have been a peak CO_2 concentration of over 5,000 ppm about 530 million years ago without a corresponding extinction occurring. Sixth, there is a strong periodicity to past extinctions with a twenty-five to thirty-million-year cycle (there are only a couple of exceptions in the last ten cycles). Some have suggested an outside mechanism may therefore be at work. It is unclear what outside mechanism would have a cycle length of as long as thirty million years. Some scientists have postulated it could be a giant nemesis comet circling our Sun.[21] This could be supported by iridium concentration spikes found in geologic layers, presumably caused by the high iridium content of meteorites or other falling space material. The famous K-T boundary iridium spike found in several locations around the globe at sixty-five million years ago corresponds to the Cretaceous-Tertiary extinction. Seventh, a few of the past extinctions do not follow an increase in CO_2 concentration at all. This suggests the extinctions themselves may have been what caused each subsequent reduction in CO_2 concentration, seemingly indicating elevated CO_2 levels are not really the *cause* of the extinctions as some have postulated, but rather it may be an *effect* resulting from a sudden reduction in the quantity of CO_2-producing life on the planet, and the extinctions themselves are actually caused by something else entirely different. Eighth, it cannot be reasonably expected that mass extinctions will not occur again—naturally.

Therefore, the question must be asked if our expectation is to control even the natural phenomena, such as Earth's volcanic activity or orbiting comets, in order to prolong our present existence or to avoid another evolutionary cycle or phase—which may be nature's way of changing the game and improving life?

9. WHY DO SOME BELIEVE THAT GLOBAL WARMING MAY NOT NECESSARILY BE BAD?

Focus: Perceived potential positive impacts of global warming, despite the potential for catastrophic consequences, are mentioned for a balanced perspective.

Layman's Brief:

There is very little argument about the reality of a *greenhouse effect*. And most agree the Earth has also experienced *global warming* over the last century. Most agree that at least some of this global warming is human-caused. However, there are some who say global warming may not necessarily be bad, and they cite the benefits of global warming. Some of these benefits include more temperate temperatures around the globe, agricultural benefits, and even the possibility of character improvement due to increased hardship.

Several economic studies have attempted to determine the economic costs of a rise in global temperatures. Most have found that global warming is expected to have a significant negative impact. However, at least one study has found an overall economic benefit from global warming.

Quick Reference:

- Any weather pattern change, from any cause, will have a likely redistribution, and possibly increased variability, of precipitation and temperatures. These changes would obviously impact some areas and some countries much more than others. In fact, these changes in climate patterns are potentially the largest impact from global warming.
- Warming at the poles appears to be two to ten times greater than the average global temperature warming. Winter warming is about twice that of summer warming. Nighttime temperatures are rising faster than daytime temperatures.
- As with any perceived change, potential or real, there are advantages and disadvantages, gains and losses, winners and losers.
- In particular, warming would be expected to have more benefits to those living in colder climates, such as Canada. Warming could lengthen crop growing seasons, and warming has been shown to positively impact human health. Increased atmospheric CO_2 has been shown to enhance plant growth and increase the health-promoting properties of the food we eat.
- Moreover, multiple wise men have observed and stated for millennia that challenges and hardships build character in man, even if man does not like it. Any potential future hardship from global warming, or alternatively the measures required to stabilize or reverse it, could have a positive impact on man's character—which some argue is more important than health or physical life itself.
- Some consider the possibility that mankind's best hope and fate may be for society to collapse, even potentially

catastrophically, in order for man to ultimately progress past man's current problems.

- Society, as we know it today, is unsustainable in a multitude of ways.
- Mankind has arguably been on a social/character/moral decline, and this decline appears to be accelerating today.
- Man enjoys thinking he *can* control or is *in* control of his own destiny—and yet, if man does *not* really have this control, man may find he is fighting even God or, if you prefer, an inevitable course of the universe, which will occur whether we embrace it or oppose it.
- Biblically, a catastrophic collapse has occurred before; it is predicted and may even be required in the future in order for mankind to progress. So we should not be too dismayed about it.
- Change scares most of us, and radical change is even scarier. One lesson appears relatively clear: it would not harm us to be less stubborn and to improve our adaptability to any change, including a change in our climate.

More Detail:

More Temperate Temperatures

It is likely that weather patterns will indeed change due to some amount of global warming and climate changes, either naturally or by human cause. Warming at the poles appears to be two to ten times greater than the average global temperature warming. Winter warming is about twice that of summer warming. Nighttime temperatures are rising faster than daytime temperatures. With these trends, it should be expected that average weather and average temperatures might become more uniform around the globe. Many people would not consider this to be bad.

Improvements in Agriculture

Essentially all plant life can be expected to flourish more with increased atmospheric CO_2 levels. This is being confirmed in more and more studies. However, this should not be too much of a surprise. It is not uncommon for ordinary vegetable and flower greenhouses to utilize increased CO_2 levels by pumping in extra CO_2 over the plants in order to spur increased plant growth. How much will this increased plant growth increase the natural CO_2 sequestration and reduce the net increase in atmospheric CO_2 concentration? Scientists don't really know for sure, and this effect is currently not modeled by GCMs. It can be expected there will be winners and losers in any change, but historically colder regions and regions closer to the poles should have longer growing seasons with increased global warming. Canada, northern Russia, and even northern Europe could see greater agricultural production in the future, and these regions might even emerge as great agricultural powerhouses.

The Economic Impact of Global Warming

The Intergovernmental Panel on Climate Change found that the costs of global warming would range from 1.5 to 2% of GDP for the world and about 1 to 2% for the U.S. Yet, Fred Singer, an atmospheric scientist and president of the Science and Environmental Policy Project, notes that several sectors were not included in the IPCC analysis and non-market sectors all suffered from negative impacts, and much of the time the impacts were greater than in the market sectors. Since no reliable metric was agreed upon to measure non-market impacts, the IPCC results are little more than an assumption by the IPCC.

Robert Mendelsohn at the Yale School of Forestry and Environmental Studies and James E. Neumann with Industrialized Economics Inc. found a different result than the IPCC.[1] To start with, Mendelsohn and Neumann assume a doubling of CO_2 that

STEVEN E. SONDERGARD

would lead to a 2.5°C increase in global temperatures. Note that this temperature increase is probably on the high side. They also include sectors of the economy that were ignored by the IPCC, such as commercial fishing. Their allegedly improved analysis also considers the possibility of human adaptation.

Mendelsohn and Neumann found that overall the economic impact of global warming on the U.S. might be positive, with about a 0.2% increase in GDP. This includes large positive impacts on agriculture and smaller positive impacts on forestry and recreation. All other sectors might experience negative impacts, but far smaller than what was found by the IPCC. There were twenty-six economists involved with the writing and reviewing of their work. Not only were individual chapters reviewed, but also several economists reviewed the overall work. A review of their work can be found at http://www.sepp.com.

Another aspect that needs to be carefully addressed relates to the outcomes that would occur whether anthropogenic Global Warming occurs or not. Climate cannot be expected to remain constant naturally, and any natural impacts on mankind should logically be subtracted from the anthropogenic impacts on mankind. It appears that a reasonable assessment of the natural impacts has not been done, and proper accounting for these economic impacts has not been done either. Properly accounting for natural impacts would necessarily alter the quantification of anthropogenic impacts.

Hardships Can Be Beneficial

Any pending catastrophe from global warming could obviously induce some magnitude of increased hardship for man. A counter-argument could be made that any increase in hardship could result in improved character for man. Most would not dismiss that there is at least some value with this. Biblically, character is considered of even more value than either life or health, both of which are commonly held in higher esteem by modern society.

It is not suggested here that a catastrophic collapse is the only future outcome, or even that it is the most probable outcome. But societal collapse *is* a strong candidate for a future scenario, and it must be examined and considered. Still, if a catastrophic collapse does occur, there can nevertheless be hope for mankind. In fact, some might argue that this outcome may provide more ultimate hope for mankind than less extreme outcomes.

Change scares most of us, particularly uncontrolled external change, including a change of the magnitude we could observe from this cause. Yet, mankind cannot control everything, even if we wanted to. We should, however, strive to sensibly and reasonably control what we can. One of the things over which we still have control is the change within ourselves. We still have the choice to choose to control ourselves and our individual habits, and we can prepare to adapt to the change we cannot control.

STEVEN E. SONDERGARD

10. CAN MANKIND TAKE ACTION TO AVOID A CLIMATE "JUMP" OR AVOID THE NEXT ICE AGE?

Focus: Before this question is answered, consideration is given to the long-term perspective and what proposed climate mechanisms are available.

Layman's Brief:

The ice core record indicates our past climate has been wildly variable, naturally. This alone says abrupt climate changes can be expected in the future. The current interglacial has been a period of stable climate relative to most of our climate history, and the current interglacial is longer than most of the known past interglacial periods. The anthropogenic impacts on climate must be overlaid over this climate history to ascertain a probable future scenario.

Many have concluded that by making choices now to reduce

global GHG emissions we can slow the rate of global warming and reduce the likelihood of undesirable climate changes. One mechanism of an abrupt climate change has been proposed that could shut down ocean currents and plunge the Earth into an ice age. Yet, most scientists are rejecting this mechanism today.

There is other evidence that increasing atmospheric greenhouse gases could delay the next ice age, or an abrupt climate change, even by thousands of years.

Quick Reference:

- Scientists do not yet understand all of the cause-and-effect relationships required to truly determine if a climate jump could be avoided. However, there is probably enough data for a realistic educated guess to be made.
- As previously indicated, the timing of major ice ages is determined by the solar variations resulting from Earth's natural near-circular to more-elliptical orbital variances, Earth inclination changes, and Earth wobbling. The varying solar energy output from our Sun will also have impact.
- Based on both our Earth's known orbital variations and ice core analyses, our Earth would be expected to have another natural periodic ice age—approached gradually but jumpy over the next ninety thousand years.
- Our Earth has already experienced approximately 11,000 years of relatively steady warmth, which is already greater than the approximate 10,000 years the ice core record might suggest is the norm. Our Earth might be considered currently overdue for naturally variable and jumpy climate changes to begin again.
- There is a possibility human-caused greenhouse gas emissions could trigger extra warming and ice sheet melting to change the Atlantic currents and trigger a climate jump—as has been somewhat widely promoted. But this is a possibility that appears to have a low probability. Many

STEVEN E. SONDERGARD

other possibilities seem to have higher probabilities, even possibilities with positive consequences.

- There is a reasonable possibility that increased atmospheric GHG concentrations may have already extended our time before jumpy weather patterns are resumed in our ensuing march toward the next ice age. In addition, there is a reasonable possibility global warming might actually help further delay a feared major climate jump. Indeed, this delay in a climate jump appears quite probable to some scientists.
- Ironically, avoiding the next ice age might be possible by increasing the atmospheric CO_2 concentration, and it would probably need to be increased by several multiples before it could completely offset and overcome the known natural orbital variations of our Earth.
- The global unity required and the cost to realistically manage the climate both indicate it will be unlikely that the next ice age could be avoided altogether. In addition, it appears the proposed action measures to counter Global Warming could easily be counterproductive to a more stable climate.

More Detail:

One computer model at the University of East Anglia in England suggests the increase in greenhouse gases could delay the next ice age by as much as 50,000 years. This is quite a long time, and it would extend the current interglacial to multiples of all known past interglacial periods. Even if the actual number were a fraction of this time span, it would be much longer than our civilization can reasonably be expected to last. Whatever the actual length of the delay, there is a growing camp of experts who believe increased CO_2 levels might delay the next ice age by one thousand to several thousand years. Some even suggest we would have already started the next cooling phase toward an ice age if it were not for the impact of agriculture.[1]

There is a fairly high likelihood that increased atmospheric GHG concentrations may have already extended our time before the next ice age comes. In addition, there is a reasonable possibility that global warming might actually help further delay a feared major climate jump. Indeed, this delayed impact appears quite probable to some scientists because the warming could reduce and delay the impact of the Milankovitch cycles and extend the time before the next ice age ensues or climate gyrations resume.

There is another camp of experts who suggest increased CO_2 levels may actually trigger the next ice age by shutting down the ocean current circuit. This mechanism was first forwarded by Wally Broecker in the 1990s. One description of this mechanism can be found in *Under a Green Sky* by Peter Ward, 2007.[2] The result would forecast a paradoxical flip-flop in which global warming causes a major cold spell. However, this outcome appears unlikely because the impact would need to be relatively sudden, such as past catastrophic ice dam collapses resulting in sudden ice-melt flooding. In addition, previous warm periods were warmer and lasted longer than the current warming trend without observing this flip-flop. Furthermore, the concept is fundamentally flawed in that the temperature would need to rise by maybe 7°F to 9°F in order to flip the switch, and then the subsequent cold spell would not create temperatures much colder than prior to the temperature rise. This position gained popularity just a few years ago, but it is finding fewer followers today as scientists have analyzed the concept in more detail, and Wally Broecker himself has reversed his stance.

Our past indicators of climate also suggest more variability in our climate should be naturally expected than what modern man has experienced over the last 10,000 years. Extrapolating these past indicators to our present situation implies that our climate could start sliding toward change or the climate could suddenly jump at any time—naturally.

Virtually all meteorological textbooks proclaim weather is predominantly driven by the temperature differences between the equator and the poles. The increase in greenhouse gases has the effect of warming nighttime temperatures more than daytime temperatures and warming the temperatures near the poles relative to temperatures near the equator. This leads us to a conclusion that increased greenhouse gases might have the general impact of reducing the variability in weather, and it could potentially reduce abrupt climate changes. This is diametrically opposed to many reports indicating the effect of high greenhouse gas levels would increase the likelihood of climate jumps.

The Milankovitch orbital variations are driven by solar system planetary forces. These forces in turn drive the glacial and interglacial cycles on Earth, and they appear to be far too great for man to be able to completely compensate for them. Most would agree these planetary forces are beyond our direct control. That leaves only responsive options such as a mechanism that would counteract the planetary forces in some manner. Ironically, there is a reasonable possibility that if atmospheric CO_2 concentrations could be increased by several multiples it might be possible to offset and overcome the known natural orbital variations of our Earth and prevent or at least diminish the next ice age. That appears to be what occurred over most of the last 600 million years prior to the ice ages. This possibility leads to the conclusion that the proposed action measures to counter Global Warming could be counterproductive to a more stable climate.

Increasing CO_2 concentrations might be tempered and managed, if needed, by taking action measures that cool the atmosphere. This might include the addition of certain atmospheric aerosols that don't have long atmospheric lifetimes. In fact, the direct and indirect impacts from aerosols are reported by the IPCC to have a combined forcing of -1.2 W/m^2, with a range of -2.7 to -0.4 W/m^2. This impact is relatively large. However, it should be obvious that an attempt to manage global climate to this degree would be risky. There would likely be an outcry concerning the potential downside to human health in proceeding with this action. To minimize the risk, it would be best if scientists understood all the natural forces at work including the magnitude, timing, duration, and effect of these forces. Full knowledge would also be recommended of the impacts of any selected countermeasures and their side effects. In addition, the global unity and the cost to pursue an effort such as this would make it very improbable. It therefore appears unlikely that humans would be able to completely counter and defer the inevitable and impending natural climate swings.

It is concluded that it might be possible, but it is very unlikely, that the next ice age could be avoided altogether. And some would argue that humans should not desire or attempt this manipulative ability, even if we believe we could achieve it—as this would wield too much power, and man always seems to incur unintended consequences when he wields too much power. With this conclusion, many contend that humans should simply be more adaptable.

STEVEN E. SONDERGARD

11. HOW MUCH SHOULD HUMAN-CAUSED GREENHOUSE GAS PRODUCTION BE REDUCED?

Focus: Discussion of the Kyoto protocol, the Stern Report—with comments, the IPCC reports (Working Groups I, II, and III)—with comments, Europe's Emissions Trading Scheme, the expected patchwork of regulations, the USA Lieberman-Warner Climate Change Bill, and the California-initiated Low Carbon Fuel Standard.

Layman's Brief:

Many large ecological and societal consequences are projected from global warming, and many suggest these consequences should be minimized. It is believed by many that erring on the side of over-reaction is warranted to preserve our future options and allow further technical solutions to be developed. Rapid global average temperature increases are projected to lead to many dire outcomes. The scale of this challenge is immense, and industrialized countries are expected to start immediately down a path of deep GHG emission reductions. It is hoped that undeveloped countries would follow the lead of industrialized countries. It is portrayed that a global reduction in emission quantities should be roughly 60% to 80% of the emissions relative to today. Any delay in taking action is proclaimed to almost certainly result in greater costs over the long term.

On the other hand, the rapid temperature increases are proclaimed by others to be either overstated or mostly natural in origin. The forecasted outcomes are decried to be exaggerated with many outcomes expected to occur from natural causes. Industrialized countries who initiate aggressive mitigation measures would incur high economic penalties. Performing measures in developing countries would have a higher benefit-to-cost ratio than in industrialized countries, and it is therefore important that developing countries participate in any action. The arguments continue that it would be unprecedented and unlikely for political alignment to occur, and those who pursue aggressive action first or without alignment would be at further disadvantage. Even with alignment, the economic penalties of extensive mitigation measures are proclaimed to be severe. Uncertainty about causes of global warming and the high downside of overreaction call for proceeding with caution. Only the pursuit of lower-impact countermeasures is warranted. Overreaction would take away attention and resources from other more-pressing societal issues.

Quick Reference:

- The main arguments are: 1) We should do what we can, even if it is small. 2) We should try to avoid rapid changes in greenhouse gases that might trigger a premature jump in our climate, prior to nature triggering a climate jump for us. 3) We should err on the side of overreaction to have the best chance at saving the planet, and 4) It will not cost that much.

STEVEN E. SONDERGARD

- The main counterarguments are: 1) Man has only a small impact on total GHG production. 2) Many forecasted outcomes are expected to occur anyway. 3) The total solar effect minimizes the impact of increased CO_2 levels on global warming. 4) Anthropogenic Global Warming appears less severe than other societal impacts and less severe than the negative impacts of vigorously reducing greenhouse gases, and 5) it will have a high cost.
- But to answer the basic question of this chapter, we should ask another basic question: can mankind reverse or control global warming, or are we simply being arrogant? Then, if we do have control, do we have the global unity and the intestinal fortitude to do something about it? These questions are more pointed, but they get closer to the root of the answer.
- When the social and economic consequences of substantially reducing man-made greenhouse gases are considered, they can arguably be more severe than the estimated benefit obtained by this action, particularly in industrialized countries.
- Alternatively, the Stern Report, 2006, projects these social and economic consequences of reducing greenhouse gases to be relatively minor and the reductions in CO_2 to be manageable—which is described as a 60% to 80% absolute reduction in *all* greenhouse gas emissions by 2050, and with a worldwide power sector that is 60% decarbonized.
- A closer examination of this Stern Report directive reveals that for North America it would be like shutting down *all* industrial, transportation, and commercial sources. Human breathing, waste generation, and agriculture could remain. These are startlingly extreme changes in how mankind would live—even if we were to proceed just halfway towards that goal. Virtually all aggressive plans stipulate stopping and reversing the modernization and economic prosperity mechanisms of all industrialized countries. Even more important to successfully reducing global emissions is stopping the growth and corresponding emission increases from developing countries. It

is increasingly apparent that most effective plans will also require either a stabilization or reduction in human population.

- The IPCC AR4 action plan of May 2007 recommends a plan for annual emissions to peak by 2015 at 10% to 20% higher than today's levels then fall to 50% to 85% of 2000 levels by 2050. Even this target is projected to cost multi-trillions of dollars and is being dismissed by many world political leaders as unrealistic. They fear it would cause a worldwide economic recession or depression.

- The Kyoto protocol, China, and most developing countries each argue that *industrialized* nations should bear the brunt of reversing global warming. Even though China is already the world's second largest GHG emitter, the largest CO_2 emitter, and has been starting up almost two new 500 MW coal-fired power plants every week, they are not considered an industrialized country. On the other hand, the UN and the latest IPCC report conclude the greatest potential for curbing future Global Warming lies in the *developing* nations, since those are the countries that tend to be the least efficient and that are now growing the fastest. This debate will likely continue for some time.

- In May 2007 the tide appeared to be changing when President Bush positioned the U.S. to act—if developing countries would also comply. President Bush then called an environmental summit in late September 2007 to discuss coordinated reduction plans. This meeting did not yield much action. The 2008 G8 Summit in Japan yielded agreement that industrialized countries would work together to reduce carbon emissions to 50% by 2050, but the baseline year was not established. There was also a prerequisite that major developing countries must also participate.

- China, currently the world's second largest GHG emitter, is expected to overtake the U.S. in GHG emissions between 2008 and 2010; and India, currently the fourth largest emitter, is rapidly industrializing as well. Neither country is required to control GHG emissions per the Kyoto Protocol. In addition, Brazil (number ten), South Korea (number

STEVEN E. SONDERGARD

eleven), Mexico (number thirteen), Indonesia (number fifteen), and a total of over 140 developing countries have no specific scheduled emission commitments under the Kyoto Protocol. If this type of disparity continues when laws are promulgated, the obvious consequence is that emitting companies will shut down and new industry will resurrect in countries where they would not be required to pay for their emissions. This is already happening. Industrial production has been moving to China and India. This transition could be expected to accelerate further, and overall global emissions would not diminish but probably increase.

- Carbon trading and carbon credits essentially transfer wealth from industrialized nations or areas to developing nations or areas. It is a mechanism whereby developing nations can catch up with industrialized nations. The cost of carbon today in Europe's Emissions Trading Scheme bounces around $20/tonne to $25/tonne, and the IPCC projected in May 2007 that the cost of carbon emissions would reach $100/ton by 2030. In addition, Congressman Dingell (D-Mich) recently (August 2007) proposed a $100/ton carbon tax (later changed to a $50/ton tax in September 2007), plus a 50¢ per gallon gasoline tax increase.

- In January 2007, California announced the Low Carbon Fuel Standard (LCFS) that will require a 10% reduction by 2020 in the GHG emissions that a fuel emits over its life cycle (well-to-wheels, if you will). Other U.S. states and three Canadian provinces have already committed to joining this stance. The EU and Japan are watching closely. This could have far reaching implications on oil sands and heavy crude production and pricing. The biggest challenge will be in how this will be measured and exactly how policies would work.

- The Earth's atmospheric conditions encompass us all. Regional-only laws and actions to reduce greenhouse gases cannot ultimately be successful. Yet, that is what is happening today. Whatever the plan to reduce Global Warming, it will likely need to have global alignment approaching unanimity. This would require most of the world's disparate factions to perceive that the probable upside—for them—of the proposed measures required to reverse Global Warming are indeed better than the probable downside of Global Warming itself. The trade-offs that every governing entity should consider include:

1. economic competitiveness,
2. energy security, and
3. quality of life decisions.

- Individual comfort and greed will need to be overcome by aggregated fear for significant action measures to be enacted globally. For this to occur, the logic of cause-and-effect relationships should be rock solid and the outcome assuredly bad—both of which are not quite the case at the moment. Therefore, to spur action more quickly, it is likely that an increased number of projections will be given about Global Warming that are designed to instill fear, whether they are valid or not.

- The Lieberman-Warner bill was designed to reduce U.S. GHG emissions by 63% to 71% of 2005 levels by 2050 using a cap-and-trade system. If this or a similar bill is enacted, it would be the toughest global warming bill in the world. The biggest fear is that a bill such as this would do significant harm to the U.S. economy.

- A successfully coordinated and implemented worldwide plan, involving most nations, of the magnitude required to reduce global warming, if this is required, would be unprecedented in human history and appears quite unlikely. The effective abandonment of the Kyoto protocol by most nations that signed it is evidence of how difficult a unified effort will be.

- An objective evaluation suggests some of the more vigorous action measures may simply have unintended consequences and higher risks than are currently perceived. These unintended consequences plus projected consequences lead to difficult trade-offs between Global Warming, our economy,

STEVEN E. SONDERGARD

energy security, and our quality of life. How much we should reduce the impact of Global Warming must be determined in relation to other societal impacts and by examining the trade-offs between every one of these impacts.

- Considering all the perspectives, a climate change plan that embraces less forceful action appears most prudent. Plans that propose aggressive action would be very difficult to achieve and/or be counterproductive to the cause or to society as a whole. The costs of further action measures also appear to be quite excessive, particularly in relation to the benefit. After analyzing the probable upsides and probable downsides to Global Warming, based on establishing their probable risks and probable consequences, taking aggressive action is actually discouraged.

More Detail:

Global Greenhouse Gas Emissions Breakdown in Year 2000

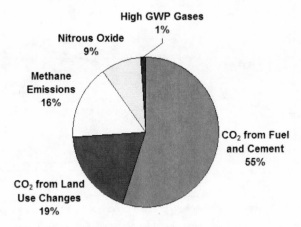

Figure 11.1. Global greenhouse gas emissions in year 2000. (Adapted from: Methane to Markets Partnership Fact Sheet and http://www.epa.gov/climatechange/emissions/globalghg.html)

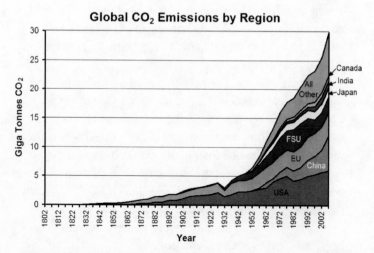

Figure 11.2. Major global sources of CO_2 emissions by country. Note the rate at which China has increased emissions. (Adapted from: Carbon Dioxide Information Analysis Center and http://www.epa.gov/climatechange/emissions/globalghg.html)

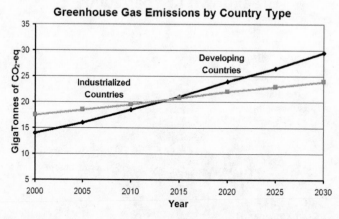

Figure 11.3. Future greenhouse gas emission projections by country type. Note that industrialized countries' emissions are relatively flat, while developing countries are increasing rapidly. (Adapted from: SGM Energy Modeling Forum EMF-21 Projections, Energy Journal Special Issue, and EPA's Global Anthropogenic Emissions of Non-CO_2 Greenhouse Gases 1990–2000, and http://www.epa.gov/climatechange/emissions/globalghg.html)

STEVEN E. SONDERGARD

The Kyoto Protocol

In December 1997, more than 160 countries came together to finalize the Kyoto protocol. It was designed to be the world's primary mechanism to combat climate change. It is viewed only as a first step by proponents. The Kyoto Protocol established targets and timetables for cutting the emissions of six greenhouse gases or groups of gases for Annex 1 countries, which are industrialized countries. These Annex 1 countries include the members of the Organization of Economic Cooperation and Development (OECD) and twelve additional countries from the former Soviet Union (FSU) in eastern and central Europe. Emission targets for each country are based on each country's 1990 CO_2 emissions level. Little consideration was given to the cost of compliance or the impact on economic growth. Each country is required to submit an annual GHG inventory.

It was not until February 2004 that the Kyoto Protocol entered into force, ninety days following the ratification by Canada and then Russia, which triggered the fifty-five nations and the 55% hurdle requirement for Annex 1 CO_2 emissions, as stipulated in article twenty-five of the protocol. There are currently more than 180 countries that have signed the Kyoto Protocol. The governments of the U.S. and Australia never ratified the Kyoto Protocol because of the potentially negative impacts of the emission reductions for each country.

Non-Annex 1 countries, sometimes called Annex 2 parties, are developing countries that have signed the protocol but are not obligated to have targets and timetables for emission reduction. These Annex 2 parties include the Middle Eastern countries of Iran, Kuwait, Qatar, Saudi Arabia, and the United Arab Emirates. Other countries that do not have emission reduction targets include China, India, Brazil, Mexico, Indonesia, and South Korea.

There are three mechanisms established by the protocol for crediting emission reductions. These mechanisms are: 1) emissions trading, 2) joint implementation (JI) projects, and 3) clean development mechanism (CDM) projects. Emission trading allows industrialized countries to trade emission credits among themselves. Joint implementation projects allow cost sharing between Annex 1 countries. CDM projects allow investment in developing countries to offset emissions from Annex 1 countries through certified emission reductions (CERs). The global market for CDM trading in 2007 was about $50 billion.

With the Kyoto Protocol being considered only a first step, it is set to expire in 2012. Formal negotiations were held in December 2007 on a successor agreement at a conference in Bali, Indonesia. An irony is that some recent (2007) calculations now put Indonesia as the third largest GHG emitter in the world, due in large part to significant deforestation for economic growth, relative to their 1990 baseline. As this book was going into production in December 2008, the UN Climate Change Conference took place in Poland, in which framework targets were set for a 20% reduction in GHGs, 20% renewable energy, and 20% reduction in energy used by 2020. A new global agreement based on these negotiations is targeted and expected to be established in December 2009 in Copenhagen.

Opposing Comments on the Kyoto Protocol

The Kyoto Protocol does not consider risks or costs in its action guidelines, yet it does consider the risks and costs of inaction as infeasible. The cost of a fully implemented Kyoto Protocol has subsequently been estimated to be about $180 billion *per year*.[1] Interestingly, the impact that the protocol would have on climate is commonly agreed to be negligible, even if the protocol were 100% successful. The estimated benefit from Kyoto is a reduction in emissions of less than 1%.[2] Using IPCC temperature sensitivities, the resulting temperature impact of a perfectly implemented Kyoto Protocol would be 0.1°F lower in 2050 and 0.3°F lower in 2100.[3] If every country implemented Kyoto and complied with the agreement throughout the twenty-first century, the projected rise in sea level would be postponed by only about four years.[4] It has been estimated that hurricane damage could be reduced by about half of a percent with a fully implemented Kyoto, and the risk of malaria could be reduced by 0.2% in eighty years.[5,6] The bottom line is that for every dollar spent on mitigation efforts, the global benefit would be only about 34¢.[7] The original hope was that the policies would slow down global warming and the benefits would then appear. Even then, the implementation costs were considered high. Since any initial efforts should have the best cost-benefit ratio, if Kyoto fails this simple cost-benefit test, then all subsequent efforts should have inadequate benefits for their costs as well.

Most industrialized countries are already having difficulty meeting their emissions reduction targets. In August 2007, Canada had already declared defeat in a Canadian environmental ministry report with 27% increase in emissions so far, rather than the prescribed reduction. In 2003, New Zealand emissions were up 14%, Japan's emissions were up 9%, the UK was up 4%, and Western Europe was either up 2% or slightly down. By 2006, Japan was up another 4% to 13%, the U.S. was reported up 16%, and the European Union (without eastern European countries) was reported down 2.7%, including credits. Many contend that the prescribed Kyoto reductions could only be met with a global recession or depression. In 2005, after considerable UK effort, Tony Blair, the British prime minister at the time, stated that no country will want to sacrifice its economy in order to meet this challenge.

The 1990 baseline established by the protocol has proven to be more advantageous to some countries compared to others. Virtually all of the countries that are reportedly meeting their targets had an advantage of some kind, such as Russia and eastern European countries, with their 1990 baseline relative to other countries. For example, Germany had the advantage of being combined with a poor starting baseline from East Germany prior to the collapse of the Berlin wall, and by 1997 they already had a 9% advantage. Other European Union countries, such as Great Britain, had the advantage of a coal burning heritage. All of the nations or regions that have reduced their emissions by the largest amounts are former Eastern Bloc countries that have undergone drastic economic declines since 1990. The EU plans to meet its obligation to Kyoto by distributing the benefits among all its members. It can be debated as to whether this is fair, but the reality is that an uneven playing field can hinder actual emission reductions. Overall European emissions are not falling, and the only way the EU will meet Kyoto is via CDM projects. This is a key point. Further, evidence exists that many carbon credits have been given to projects that were already completed or would have been done anyway.

The U.S. is often criticized for not ratifying the Kyoto Protocol. Yet, the U.S. would have had the highest cost of compliance (possibly in the trillions of dollars) and with negligible results.[8] In addition, Canada and Australia would also have very large costs for compliance, with very little benefits. It is no wonder the U.S. and Australia did not sign it. Canada's prime minister signed it late, and only after he knew he was leaving his office to another political party.

China and the U.S. are the largest emitters of both CO_2 and GHG with combined emissions at about 40% of the world total. (India is currently the fifth largest emitter of CO_2 and headed to be the third largest emitter of CO_2 by 2030.) The U.S. is the largest emitter of GHG, and China just surpassed (2007) the U.S. as the largest emitter of CO_2. It is often stated that the U.S. needs to initiate action first if China is ever expected to take any steps to reduce emissions. That may be true as far as it goes. Yet, it is unlikely that China will significantly change their growth path in order to control emissions, even if other countries were to take action first. China conveniently calculates its GHG emissions per capita, and it unapologetically declares its priority is to grow first and let the wealthier nations (again calculated per capita) lead the effort and clean up the mess they created. There appears to be little consideration by China that it may be contributing significantly to the overall global problem itself.

China had a 2001 news conference where they were praised for reducing emissions by 14% within four years. However, within a year it was determined that these official numbers were bogus. Further, China has continued to manufacture HFC-23 that the rest of the world banned decades ago. Recently China has been getting paid (via Kyoto CDM projects) to shut down these plants as offsets for other projects in industrialized countries around the world. Chinese industry has found that their HFC-23 production can be more valuable than any other products they produce. There is so much money being made on these plant shutdowns that China has imposed a 65% tax on these revenues, and China itself is making billions. China will continue to seek offset payments for cleaning up dirty businesses in the future. Obviously, this practice is controversial. The Chinese can be expected to deliver rhetoric and enhance their image as they position and posture themselves on the world stage. But China's culture precludes them from taking action that does not benefit China, particularly with respect to the rest of the world. In fact, some have calculated that global warming would be a net benefit for China and that China would lose fewer lives to less-severe winters than to warmer summers. Chinese cooperation solely for the benefit of humanity should not be counted on until actual observations confirm it.

It appears unlikely that China would ever take any meaningful action on real GHG emissions reduction, unless it was strategic or economic for itself. It never hurts to try of course, but to expect a different outcome appears unrealistic. China's growth engine must continue or it will create large internal social issues of its own. The magnitude of Chinese growth is staggering. Coal is predominantly what drives China's economic engine. China has built almost two large coal-fired power plants every week for the past five years, and it is projected to build one per week in the coming five years. China is building vast middle class housing projects with cheap labor. It is projected that by 2015 China will consume 50% of the world's copper demand. Steel and other commodities have a similar trajectory. Therefore, future

Chinese emissions relative to 1990, or to any other benchmark date, are not realistically expected to go down anytime soon.

The Stern Report

The Stern Report has been widely distributed and is often cited.[9] For this reason alone it has significance. This report will first be summarized, and then some comments will be made.

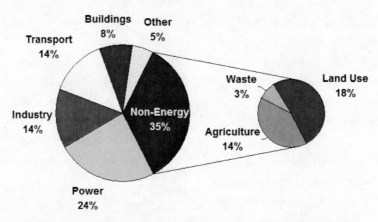

Stern Global GHG Emissions in Year 2000

Total GHG Emissions in Year 2000 were 42 GigaTonnes of CO_2-eq

Buildings 8%
Other 5%
Transport 14%
Industry 14%
Non-Energy 35%
Power 24%
Waste 3%
Land Use 18%
Agriculture 14%

Figure 11.4. Breakdown of greenhouse gas emissions in year 2000. (Adapted from: Stern Report, Executive Summary, p. iv; http://www.hm-treasury.gov.uk/media/4/3/Executive_Summary.pdf, using data drawn from World Resources Institute Climate analysis Indicators Tool (CAIT) online database version 3.0)

The Stern Report was completed in 2006 by a team led by Nicholas Stern, an economic advisor to the U.K. The report was commissioned by the Chancellor of the Exchequer reporting to both the chancellor and to the prime minister. The report starts by making the claim that "given that climate change is happening, measures to help people adapt to it are essential." Further, "climate change presents very serious global risks, and it demands an urgent global response."

The report first "examines the evidence on the economic impacts of climate change itself" then "explores the economics of stabilizing greenhouse gases." The Stern team tries to take an international perspective and stresses collective international action and cooperation. A long lead time before realizing an impact from any action is expected. The costs of mitigation must be viewed as an investment. Plus, the report claims there are opportunities for growth and development along the way. Future policy must promote sound market signals, via clarity and predictability of the rules and with little fear of policy reversal. The private sector will respond—given the right incentives.

The report considers the cost of the impacts and the cost and benefits of mitigation in three ways:

1. consideration of the physical impacts of climate change on the economy, human life, and the environment;
2. using economic models; and
3. comparisons of the incremental social costs of increased GHG emissions against the incremental abatement costs.

The report comes to a simple conclusion: that the economic benefits of taking action against climate change are strong and outweigh the costs. Early action is also described as much better than delayed action. The report claims the evidence shows that ignoring climate change will eventually damage GDP by 20% or more. Further, tackling climate change can be done in a way that does not limit the aspirations of countries, rich or poor. Plus, the earlier in time that action is taken, the less costly it will be.

The Stern Report further asserts that there are increasing risks of serious, irreversible impacts. Temperature change probabilities are attached to several impacts. Dynamic feedbacks "strongly amplify the underlying physical processes." Even if the emission levels did not increase further, the concentration of greenhouse gases in the atmosphere would increase to 550 ppm CO_2-eq by 2050 and continue beyond that into the future.

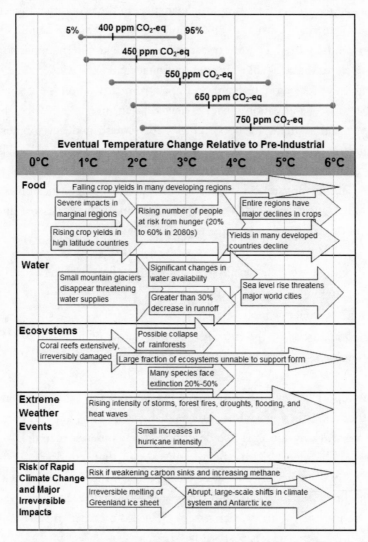

Figure 11.5. Increased atmospheric CO_2 levels and their corresponding stable temperature ranges plus an illustration of projections of various impacts for various warmings. (Adapted from: Stern Report, Executive Summary, p. v and http://www.hm-treasury.gov.uk/media/4/3/Executive_Summary.pdf)

STEVEN E. SONDERGARD

"Climate change threatens the basic elements of life for people around the world—access to water, food production, health, and use of land and the environment." Warming will have many severe physical impacts:

- Melting glaciers will increase flood risk, reduce water supplies, and threaten homes.
- Declining crop yields will occur at atmospheric temperature increases of 4°C (~7°F).
- Worldwide deaths will increase from malnutrition and heat stress.
- Rising sea levels, heavier floods, and more intense droughts will result.
- Ecosystems, including rainforests, will be vulnerable to climate change, with a startling 15%–40% of all species potentially facing extinction after only 2°C temperature rise.

"The damages from climate change will accelerate as the world gets warmer." Warming may also induce sudden shifts in regional weather patterns. Business-as-usual trends in emissions may result in temperature increases of greater than 2° to 3°C by the end of this century. (Of course, business-as-usual itself is subject to definition and assumptions.) "The temperatures that may result from unabated climate change will take the world outside the range of human experience." To avoid this, global emissions would need to be about 25% below current emission levels, and this will need to occur in the context of a global GDP that will be maybe three to four times larger than what it is today. This will be a major challenge. Emissions can be cut in four ways:

- Reducing demand
- Increasing efficiency
- Avoiding deforestation and other non-energy emissions
- Switching to low-carbon technologies

Energy efficiency has the largest potential for emissions savings. Cutting waste can often save money as well. Non-energy emissions make up one-third of total global GHG emissions, and actions to prevent further deforestation would be relatively cheap. Radical emissions cuts will be required from the power and industrial sectors, with the power sector needing to be at least 60% to 75% decarbonized. Carbon capture and storage will be essential. Deep cuts in the transport sector will likely be more difficult but ultimately needed. A portfolio of technologies will be required across sectors, and this will need to take place even with the continued economic supply of fossil fuels. Policies to reduce emissions should be based on three elements:

- Carbon pricing
- Technology policy
- Removal of barriers to behavioral change

The report emphasizes strong and immediate action but indicates policies should provide the ability to adapt. Uncertainty in the risks is an argument for more demanding goals rather than less. Carbon pricing can be initiated through taxes, trading, or regulation. Carbon pricing policy should also consider the circumstances of each country. But putting a cost on carbon emissions will not, by itself, be sufficient to bring all the necessary innovation. Even though low-carbon technologies are currently more expensive than fossil fuel alternatives, their costs are expected to fall with scale and experience. The report proposes a doubling of public investments in research and development to around $20 billion per year globally. In addition, development incentives should increase by two to five times from the current level of about $34 billion per year. Even where economics are favorable to reduce emissions, there may be behavioral barriers that prevent action. Minimum standards may be required for buildings and appliances. Informational labeling of energy use may be required to help consumers. Financial measures to assist

STEVEN E. SONDERGARD

with upfront compliance costs would help. Land use planning and performance standards would help with change. A financial safety net for the poorest in society may also be required.

The Stern Report claims previous economic modeling of climate change has been too optimistic and the reality could be much worse. The report claims it has also been relatively conservative in assessing the risks and further explains why results could be up to 20% worse than its conclusions. One particular economic model used two data sets: one set of data from the 2001 IPCC report and one set of data with slightly amplified feedbacks in the climate system. The amplified feedback data set showed 5% to 7% greater economic impact. The report makes the caveat that economic forecasting is very difficult and imprecise, particularly over fifty- to one-hundred-year time spans.

The projected temperature increase of another 2°C to 3°C by the end of this century translates into a permanent loss of 0% to 3% in global world output. The cost of increased extreme weather alone could reach 0.5% to 1% of world GDP. Heat waves will become commonplace. A 5% or 10% increase in hurricane wind speed is predicted, doubling the annual cost of damages. Flood losses in the UK alone could increase from 0.1% today to 0.2% to 0.4% with an increase of 3°C or 4°C in temperature. Climate change is a particularly grave threat to the undeveloped world, further reducing low incomes and increasing illness.

In the past, CO_2 emissions have been strongly correlated to GDP. Future reductions depend on breaking that correlation. In addition, most future emissions growth will come from today's developing countries, who want to grow their economies as inexpensively as possible. Developing countries are also struggling to improve their poverty rates, their competitive ability, and sometimes even their energy security. Yet, despite the historical pattern, the Stern Report asserts the world does not need to choose between climate change and promoting development. Hope lies in the fact that industrialized countries have at least reduced their correlations between emissions and growth.

The extrapolated forecast is that it will be possible to decarbonize our societies on the scale required to achieve both tasks.

The cost of carbon emissions will start in the range of $25 to $30 per tonne (metric ton), and the cost of carbon will also rise steadily over time. If we remain on a business-as-usual trajectory, the social cost of carbon will be of the order of $85 per tonne of CO_2. Benefits could, however, outweigh costs. Markets for low-carbon energy products are likely to be worth at least $500 billion per year by 2050. If we implement strong mitigation policies now, the net global benefits would be of the order of $2.5 trillion net present value (NPV).

International collective action will be imperative. A shared global perspective and cooperative effort toward long-term goals will be essential. The emissions of many countries are small relative to the global total and required improvement. A policy that avoids a free ride for these countries will be needed. Compensation from the international community should help. Low carbon investments in developing countries are likely to be at least $20–$30 billion per year.

The Stern Report projects that the social and economic consequences of reducing greenhouse gases should be "relatively minor" and the reductions in CO_2 should be "manageable," even when this is described as a 60% to 80% absolute reduction from 1990 in "all greenhouse gas emissions" by 2050 and with a worldwide power sector that is 60% to 75% "decarbonized." The report claims that only about 1% of global GDP per year would be required to achieve this reduction and that global stabilization at 550 ppm of CO_2-eq would result.

Opposing Comments on the Stern Report

The Stern Report detractors contend that the report includes questionable assumptions, overly menacing forecasts, and behavioral miscalculations. First, the Stern Report takes the 550 ppm CO_2-eq threshold and correlates it with a 2°C temperature increase with at least 77% to perhaps 99% probability. Then, the assumptions made about

STEVEN E. SONDERGARD

forecasted calamities of climate change appear to be very pessimistic. In addition, the effect of the mitigation efforts and the attempts to calculate their costs and benefits appears to be too optimistic. Logical strategies and policies were developed to allow mitigation, but then the assignment of probabilities and efficiencies to these strategies and policies is very rosy. The report seems idealistic rather than realistic in its action measures since perfect international collective action is considered imperative, and a perfect cooperative global effort toward consistent long-term goals is considered essential. The possibility of human adaptation is ignored in the report.

A Monte Carlo-type analysis with reasonable probability distribution assessments does not appear to have been done. The Stern Report's whole economic basis appears to be based on single-scenario models that are arguably extreme. That does not mean extreme scenarios should not be considered, because they should. However, extreme scenarios are almost never the most probable scenario particularly when dealing with large aggregated forces, and in this arena the downside consequences of overreaction are quite arguably worse than the consequences of under-reaction.

Dr. Richard S. J. Tol is a leading environmental economist. Sir Nicholas Stern utilized Dr. Richard Tol's work to reach some of his conclusions. Yet, Dr. Tol is among the Stern Report detractors. Dr. Tol has indicated the Stern Report was not peer-reviewed and suggests the damage costs for projected CO_2 levels are far too high. Many other economic and scientific authorities have also criticized the Stern Report as being a biased and alarmist political document, with inflated damages and with optimistic costs for action measures. Essentially the detractors claim that the cost-benefit analysis is incorrect and misleading.

In essence, the Stern Report takes the stance that a change in the climate would be sufficiently bad that it must be controlled, even if it were natural in origin. The predicted outcomes from various temperature increases are relatively specific within the report.

Many of the counterarguments about the report's forecasted outcomes were discussed in Chapter 8, but a synopsis follows. Our climate can be expected to change, whether caused by man or caused naturally. Ocean levels can be expected to rise simply due to natural interglacial cycles and from polar ice cap melting that has already occurred. Droughts can be expected to occur, whether they are caused by man or caused naturally. Water table levels are already being depleted, and water availability will be impacted regardless of Global Warming. Glaciers can also be expected to continue to melt from natural causes alone. Weather variability causing water shortages will vary regionally, whether caused by man or caused naturally. Famines can indeed be expected to occur, but this is much more a factor of increasing population in developing countries and poor government policy. Reduced food production appears unlikely due to human-caused Global Warming. Increased hurricane activity or wind speeds are unlikely due to human-caused Global Warming—details are described later. In addition, the forecasted outcomes, from any cause, resulting from a ~2°C rise are only guesses at best—and these guesses have very low accuracy, with the reported probabilities of many of these outcomes being quite exaggerated in the report.

The Stern Report states coordinated alignment of all disparate countries in a unified cause is essential for the long haul. There is little disagreement with that statement as far as it goes. But the language used in the report implies that this outcome would be likely, which is an idealistic stance. Remember, the magnitude of this coordinated action would be unprecedented in human history. The U.S. does not even have a reasonable, consistent, predictable energy policy. How is it expected that the world would have a unified and effective climate change policy? The global alignment appears much more problematic than what is alluded to in the report, and the probability for a uniform global policy along with effectively coordinated global actions appears extremely low.

The report states that uncertainty in the risks is an argument for more demanding goals rather than less. But this assumes the downside risks for overreaction are better than the downside risks for under-reaction. This risk assessment should, at the very least, be questioned, and on the basis of many of the counterarguments mentioned, it appears incorrect.

Stern suggests the policies against deforestation should be led by the countries where the particular forest in question stands. After all, these forests belong to these countries. There is not much dissention with that provision. The cited problem is that it will become very difficult to align each forested country against deforestation and against its own development and growth. Significant success in this area is viewed as unlikely. Deforestation in Indonesia for palm oil production is a recent example.

Carbon capture and storage is considered essential in the Stern Report. However, using projected technology coal-fired power plants are expected to cost 60% more with CCS, plus CCS operations could cost 14% more and consume up to 40% of the power generated by the plant itself.[10, 11] The infrastructure required for substantial CCS will also be quite extensive. These are very large hurdles to be overcome. They can be overcome of course—but they are not trivial and will be very expensive.

The Stern Report directive to reduce 60% to 80% of all greenhouse gas emissions does not seem "manageable" as claimed but quite extreme. This directive would be like shutting down *all* industry (14%), transportation (14%), power sources (24%), land use (18%), military activity, and forest fires! Agriculture (14%), waste generation (~5%), and human respiration (5%–10%) could remain. In the U.S. (using the EPA's breakdown), this would be like shutting down *all* industry (28%) and transportation (28%) and most commercial buildings (17%). Residential (17%), agricultural (9%), and human respiration (5%–10%) could remain. Of course, the simplistic choice to extinguish the emissions from selective sectors in this manner cannot be accomplished. For example, the residential sector will need some power and some industry just to remain effective and to be maintained. Plus, jobs will be needed from somewhere. But these simple breakdowns *do* represent the magnitude of the consequences by the proposed actions. It is apparent that the socioeconomic impact that would result from the proposed actions would be immense.

These sector breakdowns reflect startlingly extreme changes in how mankind would need to live—even if we were to proceed only halfway toward the Stern Report goal. The drastic changes perceived appear incongruent with the report's conclusion that the impact would be minimal and that the likely economic impact would only be 1% of GDP by 2050. Several major assumptions are required in order to conclude that a low impact to GDP would be possible. These include multiple technology breakthroughs, improvements in the cost of emission reduction measures, reductions in GDP being dominantly offset by GDP increases, significant sacrifices by individuals, a coordinated global effort on all fronts, and developing country participation. These assumptions in aggregate appear to be unreasonable and unrealistic.

A heavy reliance on breakthroughs in technology is a major assumption. Although some technology improvement can surely be expected, the multiple, radical, game-changing technology improvements that would be required appear to be relatively remote in the short term. Long-term coordinated action by governments, individuals, and developing countries are also major assumptions. This level of coordination would be unprecedented in human history and therefore appears very unlikely. The inefficiencies of government, greed, fear, pride, resistance to change, and lack of focused effort would all contribute to ineffectiveness.

Realistic plans to even get close to a ~70% reduction in GHG emissions would seemingly require a drastic reduction in human population. Even reasonable plans stipulate not only stopping but reversing the modernization and economic prosperity mechanism of all industrialized countries. Even more importantly, these plans require stopping the economic growth of developing countries.

The 2007 IPCC AR4 Working Group I

The Intergovernmental Panel on Climate Change (IPCC) was established in 1988 by the World Meteorological Organization (WMO) and the United Nations Environment Programme (UNEP). Every few years, the IPCC updates a report on climate change with their latest information. The first IPCC assessment report was issued in 1990. The second assessment report (SAR) was issued in 1995. The third IPCC assessment report (TAR) was issued in 2001. The three latest working groups for the fourth assessment report (AR4) each issued their reports in 2007. The IPCC's Working Group I had the task of understanding the latest physical science of climate change.[12] The report often uses the term *radiative forcing*, which is defined as the quantitative contribution to the energy balance affecting global warming, usually expressed in W/m^2. The reported findings of this working group include:

- Most of the observed increase in global average temperatures since the mid-twentieth century is very likely due to the observed increase in anthropogenic greenhouse gas concentrations. In addition, it is very likely that global climate change is not due to known natural causes alone.
- There is very high confidence that the global average net effect of human activities since 1750 has been one of warming, with a radiative forcing of +1.6 W/m^2, and with a range of +0.6 to +2.4 W/m^2.
- The combined direct and indirect radiative forcing due to anthropogenic increases in CO_2, CH_4, and N_2O is +2.3 W/m^2, with a range of +2.07 to +2.53 W/m^2.
- Human-caused aerosols produce a cooling effect with a combined direct radiative forcing of -0.5 W/m^2, with a range of -0.9 to -0.1 W/m^2, and an indirect cloud albedo

radiative forcing of -0.7 W/m², with a range of -1.8 to -0.3 W/m². Aerosols also influence cloud lifetimes and precipitation.

- Increases in tropospheric ozone contributed a forcing of +0.35 W/m², with a range of +0.25 to +0.65 W/m².
- Changes in halogenated hydrocarbons had a radiative forcing of +0.34 W/m², with a range of +0.31 to +0.37 W/m².
- Changes in surface albedo contributed a forcing of -0.2 W/m², with a range of -0.4 to 0.0 W/m².
- Changes in solar irradiance since 1750 are estimated to contribute a radiative forcing of +0.12 W/m², with a range of +0.06 to +0.30 W/m². (This estimate is less than half of the previous IPCC report, less than most reported TSI increases, and it does not include any consideration of cosmic ray impacts on low-elevation cloud formation.)
- The sea level rose by 3.0 inches from 1961 to 2003, and the total twentieth century rise in sea level is estimated to be 6.7 inches.
- The average arctic temperature increased almost twice the global average temperature in the past one hundred years, and some localized regions increased much more.
- Surprisingly, daytime and nighttime temperatures are reported to have risen at about the same rate, but this is highly variable from one region to another. It is also inconsistent with other statements in the same report declaring that nighttime temperatures are rising faster.
- There is insufficient evidence to determine trends in tornadoes, hail, lightning, or dust storms.
- Difficulties remain in simulating and attributing observed temperature changes at smaller scales.
- The global average surface warming from a doubling of atmospheric CO_2 concentrations is estimated at 3.2°C, with a range of 2.1°C to 4.4°C. This includes all feedback mechanisms.
- GCM modeling indicates that even if concentrations of greenhouse gases and aerosols could be kept constant at year 2000 levels, a further warming of about 0.2°F per

STEVEN E. SONDERGARD

decade would be expected, due mainly to the slow response of the oceans.

- Anthropogenic warming and rising sea levels, due in large part to thermal expansion, are projected to continue for centuries, due to the time scales of the various feedbacks and the ocean current circulation timeline, even with the stabilization of greenhouse gases.

Opposing Comments on Working Group I of IPCC AR4

Opponents of the IPCC contend, in general, that the IPCC reports take on an air of certainty about their conclusions, and that certainty just doesn't exist. The IPCC was established by the UN as a political body. It appears that the UN and many developing countries within the UN see the pursuit of climate change countermeasures as a means of "leveling the playing field" to help close the gap between developing countries and industrialized countries. Another problem with the IPCC reports might be traced back to the policy leaders who set the ground rules for the IPCC teams. The process drives for a consensus report and a proclaimed unanimity on subject matter that is still far from that level of certainty. The predictable result of this process requirement is that the pendulum has swung completely to one side, and most scientists with opposing views get so frustrated that they soon leave the process, or they choose not to enter the process at all. A balanced, healthy debate and evaluation of opposing views therefore never takes place. Interestingly, several participants who have opposed the IPCC conclusions are listed among the expert authors. The process and report have the appearance of a global scientific consensus. Yet, this is not the reality, as it is only the consensus of a like-minded subset of world scientists and even a subset of the authors listed.

The IPCC reports therefore take an unbalanced stance that is disconcerting. It should be apparent to those who are trying to grasp a balanced view of the subject that the IPCC reports slant their descriptions of global warming and climate change in an embellished manner. Most aspects are described with a perspective that makes it look relatively ominous. The selective sampling and the selective representation of the data is often extreme in order to amplify the issue. The reports commonly frame an aspect with limited perspective when it better suits this gloomy stance. Facts or perspectives that might support opposing or more balanced views are rarely included. It is fully understood that a great many authors contributed to the IPCC effort, but this is not a good excuse for the multiple inconsistencies within the reports, sometimes even on the same page. There are also a number of blatantly incorrect statements.

This stance by the IPCC is both unfortunate and disturbing. Good scientists do not take a "blind" stance or preclude an answer until all aspects of an analysis are examined. One example of this is the claim that the increased frequency of extreme events is simply "expected"—and then consideration is only given for supporting data. The result of this stance is a reduced trust in the objectivity of the IPCC effort. The IPCC should have access to the broadest resources and have access to the best scientific research. With this huge advantage, they should take the higher ethical road and be more responsible in their analysis and reporting as well. Many of the government authors and possibly some of the scientists appear to be so engulfed by their paradigms that it is suspected even they don't realize their own lack of objectivity. The IPCC reports are still considered by many to be the most detailed and encompassing view available on the subject. It is simply unfortunate that the IPCC reports are so far from an attempt at a balanced, honest, and intellectually mature view. It appears that an unbiased view that is both available and widely accepted is still a ways off.

STEVEN E. SONDERGARD

The 2007 IPCC AR4 Working Group II

The IPCC's Working Group II had the task of assessing the current scientific understanding of the impacts of climate change on natural, managed, and human systems.[13] It also analyzed the vulnerability of these systems and their capacity to adapt.

The key impacts of global average temperature change are grouped into five categories. These impact groups are water, food, coasts, health, and ecosystems. In general, the impacts listed in this report are less severe than the impacts listed in the Stern Report. Within the water group: there will be increased water availability in most tropics and high latitudes, decreased water availability in mid-latitudes and semi-arid low-latitudes, and more people exposed to increased water stress. Within the food group: there will be localized negative impacts, decreases in cereal productivity in low latitudes, and increases in cereal productivity in mid- to high latitudes. Within the coasts group: there will be increased damages from floods and storms and a loss of coastal wetlands. Within the health group: there will be increased burdens from malnutrition, cardio-respiratory, and infectious diseases. There will also be increased morbidity and mortality from heat waves, floods, and droughts. Within the ecosystems group: there will be an increasing risk of species extinction (at +2°C) and significant extinctions (at +4°C), coral bleaching, biosphere changes, species range shifts, wildfire risk, and ecosystem changes.

The WG-2 report does have some astute conclusions. These include the conclusion that adaptation will be necessary. In fact, the report states that more extensive adaptation will be required. Barriers, limits, and costs of adaptation will need to be overcome. The adaptation responses include technological solutions, behavioral choices, managerial practices, and policy changes. The report also points out there are formidable economic, behavioral, attitudinal, informational, social, and environmental barriers. In addition, other stresses can exacerbate our vulnerability to climate change.

The report goes on to state that the final impacts will be determined more by our vulnerabilities than by climate change itself.

Opposing Comments on IPCC AR4 Working Group II

Overall the WG-2 report was rather disappointing to some scientists. A pessimistic slant is very prominent. The Summary Report for Policymakers states: "Evidence from all continents and most oceans shows that many natural systems are being affected by regional climate changes, particularly temperature increases." This sounds pretty bad, doesn't it? But it doesn't say how big the effect might be. The effect could be infinitesimal and the statement would still be true. It can also be noted that regional climate changes would be statistically expected to occur virtually 100% of the time, even when the global climate did not change at all. Statistically, regional changes would always be expected to have increased risks of every sort. The statement doesn't even connect a cause and effect; that is, the cause isn't necessarily anthropogenic. This is an example of a statement that doesn't really say anything conclusive at all.

Another example from the report is: "Some adaptation is occurring now, to observed and projected future climate change, but on a limited basis." Unfortunately the report is chock-full of these types of statements. Some are harder to discern than others. Other statements just don't say much of significance, such as: "It is likely that anthropogenic warming has had a discernible influence on many physical and biological systems." But that is not necessarily bad—many things have discernable influences, and some of them are undoubtedly good. Still other statements made are very slanted. The report uses the wording "increased risk of extinction." Although this statement may be accurate, it is misleading as the risk of extinction could be extremely small. Whereas the term *increased risk of biological stress* would be more appropriate, it simply does not elicit the same emotional response.

Many of the report's statements about social and economic impacts are much more the result of population growth and income levels than from climate change. This is even acknowledged on p.19 of the summary report. Many other outcomes will occur, whatever mitigation actions are taken. The climate is projected to get warmer over the next century whether the GHG target levels are met and sustained or not. The same is true for increased regional weather pattern variability, sea level rise, and de-glaciation. Increased stresses and increased risks are already unavoidable due to both nature and the past. We still cannot control nature, and we cannot change the past. Even the most stringent mitigation efforts could not avoid further impacts of climate change in the next few decades. The report quantifies this impact as a further 0.6°C by the end of the century, relative to 1980–1999, even if atmospheric GHG concentrations remain at year 2000 levels. Further, most any realistic scenario projects increases in global emission levels that are very likely to be significantly higher than year 2000 levels.

In response to the impact groups of water, food, coasts, health, and ecosystems—most of the impacts from the first four groups are expected to occur no matter what mankind does in the future. Projected impacts within the ecosystems group appear to be unrealistically severe, specifically concerning species extinctions. Significant temperature changes have occurred many times in our past climate record without observing such a wave of extinctions. Plus, it must be acknowledged that biological stress can and often does make the affected species stronger.

The report does conclude that positive impacts from global warming have been observed. These observations include earlier greening of vegetation, longer growing seasons, increased crop yields (particularly at mid- to high latitudes), increased forest growth, navigable northern sea routes, and reduced demand for heating. However, even these positive benefits are given a negative spin in the summary report.

The 2007 IPCC AR4 Working Group III

The IPCC's Working Group III had the task of assessing the possible range of responses and their corresponding future emission levels.[14] The subsequently proposed action plan (May 2007) recommends annual emissions should peak by 2015 at 10% to 20% higher than today's levels then fall to 50% to 85% of 2000 levels by 2050. This target is not quite as aggressive as other proposals, such as the Stern Report, but it is still projected to cost multi-trillions of dollars and was rapidly dismissed by many world political leaders as unrealistic—as it is expected to cause worldwide economic recession or depression.

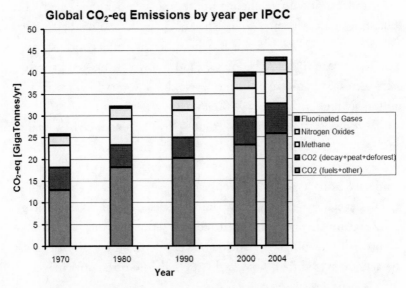

Figure 11.6. IPCC GHG emissions as CO_2-eq/year. Note that about two-thirds of the total emissions are due to CO_2. These numbers include an implied reduction for the impact of GHG interferences to obtain a more accurate overall CO_2 equivalent value. (Adapted from: 2007 IPCC AR4, Working Group III, Summary for Policymakers, p. 4)

STEVEN E. SONDERGARD

Between 1970 and 2004, annual global GHG emissions have increased 70%, from 28.7 giga tonnes (billion metric tons) to 49 giga tonnes of CO_2-eq. The largest growth of this increase is in the global power sector, with 145% increase, and most of that increase is from developing countries. During this same period of time global income growth had a 77% increase, roughly the same as the 70% increase in GHG emissions, and global population growth had almost the same increase at 69%.

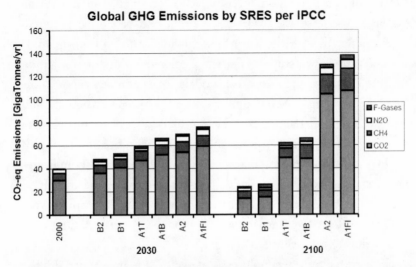

Figure 11.7. Global GHG emissions for year 2000 and projected emissions for both 2030 and 2100 from IPC SRES literature. This figure provides the emissions from the six SRES scenarios. F-gases include HFCs, PFCs, and SF_6. (Adapted from: 2007 IPCC AR4, Working Group III, Summary for Policymakers, p.8)

Future emission scenarios were examined using the IPCC Special Report on Emissions Scenarios—SRES. These scenarios are listed below:

- A1—rapid economic growth, population peaks in 2050, rapid technology improvements.
  - A1FI—fossil fuel dominates energy sources.
  - A1T—non-fossil energy sources dominate.

- A1B—balanced energy sources are utilized.
- A2—heterogeneous world, continuous population growth, non-global forces drive actions.
- B1—convergent world, same population trend as A1, service and information economy.
- B2—local solutions, continuous population growth lower than A2, some technology change.

Once potential emission levels were determined, the mitigation potential was assessed. Market potential and economic potential were evaluated, with the latter being divided into bottom-up studies and top-down studies. Perfect implementation of mitigation measures was assumed for most models. The summary report indicates this working group had much hope for mitigation actions, and with relatively little cost. But the summary also proclaims climate sensitivity is a key uncertainty about whether mitigation scenarios could meet specific temperature targets. Modeling also indicates that warming by 2030 is relatively insensitive to the specific SRES scenario and the observed temperature impact by 2030 might very likely be at least twice as large as the model-estimated natural variability during the twentieth century.[15]

Key mitigation technologies by sector are mentioned in the IPCC report, with virtually every sector requiring mitigation technologies.[16] The mitigation measures that have negative costs (positive value) have the potential to reduce global emissions by about 5 to 7 giga tonnes per year. But barriers to implementation must be overcome. The overall costs for stabilizing of CO_2-eq levels are calculated for stabilization levels between 445 ppm and 710 ppm. The cost of maintaining these stabilization levels ranges from 0.2% to <3% of global GDP. The modeling studies of the summary showed carbon prices of $20 to $80 per tonne of CO_2-eq emissions by 2030, and $30 to $155 per tonne by 2050 are consistent with stabilization at 550 ppm CO_2-eq.

STEVEN E. SONDERGARD

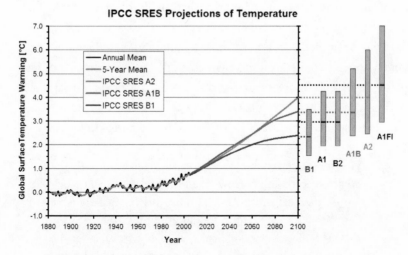

Figure 11.8. Global surface warming projections by the IPCC. This figure starts with the probability-based estimates of GHG emissions by WG-3 as reflected in Figure 11.7. These emissions were then translated into temperature increases based on the AOGCM model results of WG-1. (Adapted from: 2007 IPCC WG1-AR4, Synthesis Report, Summary for Policymakers, p.7; http://www.ncdc.noaa.gov/oa/climate/globalwarming.html)

The Working Group III summary also discusses government policy. The summary considers four criteria to evaluate policy: 1) environmental effectiveness, 2) cost effectiveness, 3) distributional effects, and 4) institutional feasibility. The integration of policies within other policies is stressed. Possible instruments of policy include regulations and standards, taxes, cap-and-trade systems, tradable permits, subsidies, tax credits, R&D funding, voluntary agreements, and awareness campaigns.

Opposing Comments on Working Group III of IPCC AR4

There is a strong correlation between GHG emissions and both income and population. These historic correlations will be difficult to break in the future. Further, the sheer scale of global fossil fuel consumption and the timeline required to change course to a different path will limit much near-term change in the trend of greenhouse gas generation. These issues impact which of the SRES scenarios might better represent future reality, and it should be obvious that each SRES scenario would not have an equal probability of occurrence. Comments on and probabilities of the SRES scenarios follow:

- A1—these scenarios may be too aggressive on both the projected economic growth and projected technology improvements. Limiting population growth is questionable if economic growth continues in developing countries.
  - A1FI—fossil fuel domination is highly probable for the near-term decades, but the best projections are that longer-term crude production will be limited or reduced.
  - A1T—the sheer scale of fossil fuel consumption will limit non-fossil fuel impact.
  - A1B—this scenario appears more likely than either A1T or A1FI due to both scale and probable crude oil forecasts, but success will depend somewhat on achieving technological breakthroughs in alternate energy sources.
- A2—at least some international cooperation can be expected to occur, with several industrialized countries taking a lead role, as long as their economies remain healthy.
- B1—service and information could indeed increase, but something far more basic must become the mechanism that powers the economic engine.
- B2—the concepts of local solutions, limited population growth, and technological advances may reflect a probable case, but this scenario may be too optimistic in their impact.

The most likely realistic single scenario, with projected high probability, is viewed to be a blended combination of A1B with some of the concepts of B2. SRES B2 alone appears to fall short on probable population growth and near-term fossil fuel use, while SRES A1B alone may underestimate the potential regulatory impact to weaken economic growth. Strong growth in developing countries is viewed as more likely, barring global economic calamity, but the expected weaker growth in industrialized countries will lower overall global growth. Analyzing the probabilities of each SRES scenario indicates that probable future growth in global GHG emissions is likely to be substantial.

The approach of Working Group III appears rather idealistic with their assumption that perfect implementation of mitigation measures throughout the twenty-first century could occur with perfect least-cost decisions, completely transparent markets, and no transaction costs. It is apparent that the working group was influenced by an idealistic thinking that disregarded several realistic probabilities and uncertainties. The summary looks at lifestyle changes with high hopes and without realistic quantification. Conversely, the history of human nature tells us that many behavioral changes would need to be externally dictated. The health benefits of projected mitigation actions are also looked at rather optimistically as opposed to realistically.

High-carbon fossil fuel alternatives, such as oil sands, heavy crude, oil shale, and coal-to-liquids, are mentioned in the report, but their impact is not assessed in the Working Group III summary. Yet, these unconventional energy sources are seen by most experts as the only way the world will come close to meeting our near-term global energy demand, and this may be even more true for North American energy security. The production of Canadian oil sands and other heavy Canadian crudes will need to increase to meet North American crude demand as other lighter crudes are either depleted or consumed within oil-producing regions.

The Working Group III Summary does state fossil fuels are projected to maintain their dominant position of the global energy source portfolio through 2030. This is a view shared by virtually all energy consultants and energy companies, and it is not just their own wishful thinking. The scale of fossil fuels is immense, and the physical and economic infrastructure built over many decades won't be easily displaced.

In addition, two-thirds to three-fourths of the forecasted global increase in CO_2 emissions are projected to come from developing, non-Annex 1 countries. This leads to the conclusion that the focus on any change in course must be primarily directed toward developing countries and that without the coordinated effort of these developing countries the world will have little chance of reducing or even limiting GHG emissions per year.

The Working Group III placed a positive spin on the potential of each mitigation measure, but the summary falls short of providing a most-probable outcome. Therefore, a non-statistically-based probable outcome is forwarded here. It is believed that a future human course can be forecasted with at least some measure of accuracy when considering basic human nature, disparate factions, cultural differences, and different driving forces. Our past does not have to dictate our future, but it usually does. It can be assumed with very high probability that the human drivers of fear and greed will not soon change. It can also be assumed that the many disparate factions of government will not suddenly become coordinated in this cause, or any other cause, for the required duration. It can be expected that China, possibly India, and probably many other developing countries will not take significant action that negatively impacts the economies of their countries. It can be assumed that economic prosperity, energy security, reducing poverty, quality of life considerations, and other driving forces will likely collide with and many times override any synchronized effort to take action against Global Warming. It can therefore be concluded that an encompassing, coordinated policy and cooperative long-term effort will be unlikely.

STEVEN E. SONDERGARD

A Patchwork of Regulations is Most Probable

Many aspects impact the best approach each country or region might take. These aspects include inherent hydrocarbon reserves, industrialized or developing status, economic development plans, current economic condition, energy security, military security, current size, current emissions, and relative greenhouse gas concern. Not every country will have the same priorities for these driving forces, nor will they have the same perspective, so it will be very hard to settle on a consensus global strategy and policy. *If* controls are universally applied, and that is a big "if," the companies who position themselves early are the ones who will flourish. Businesses might also have some influence by lobbying hard for regulatory consistency. Some companies are already taking the lead by adopting the Carbon Disclosure Project and are reporting their emissions. However, any potential universal framework will be difficult to achieve and would likely be stretched, torn, and even abandoned by some countries. A global strategy with the most hope of success will therefore be one of flexibility and trade-offs that will achieve the most benefit overall. This will still be hard to achieve, and even if it is achieved, it will assuredly upset some of the countries who see things differently.

A strong global political system currently does not exist to realistically affect the changes required. It is highly unlikely that a global climate change policy could be implemented that will govern all countries with relative equity. A non-strategic patchwork of regulations therefore appears most probable. Country, regional, and even local regulations will rise to fill the void of a coordinated global solution. Regulations at various levels are being developed extremely fast. Politicians at every level want to appear green. Politicians will jump on the bandwagon because Global Warming is in the spotlight and is popular. Politicians will likely try to outdo each other in their fervor. Differing political interests can also be expected to try to slant the playing field in their favor. A patchwork of country, state, and even local regulations will be much harder for industry and some countries to deal with. For instance, at this writing, twenty-one American states already have a renewable fuel standard. California, for one, will be requiring 20% of its power to come from renewable sources by 2017.

The European Union (EU) has already developed their Emissions Trading Scheme (ETS) to provide an effective way for EU industries to comply with carbon emission limits. Most EU companies are already purchasing and plan to further purchase credits rather than cut their own emissions. The impact of these credits is already apparent, as *actual* overall European emissions are not falling but *reported* emissions (with credits) are falling. The EU is concerned about the future ability of their companies to compete on the world stage if other countries do not regulate GHGs, and the EU is now promoting this scheme to be the basis of a worldwide emissions trading framework. Europe has already lost most of their cement plants to Northern Africa. The ETS scheme was strongly pushed by the EU countries at the December 2007 Bali conference, along with hard line caps and limits on emissions. Other industrialized countries have been concerned about derailing their own economies with specific emission reductions, particularly in relation to the huge global emissions increase from the growth of developing countries.

Developing countries are more concerned about flexibility and the ability for their countries to join the economic boom of industrialized countries. As such, they usually support flexible emission goals rather than hard targets.

Many countries are ratcheting up efficiency standards for vehicles, appliances, or industrial energy users. Voluntary programs for using less energy have also sprung up following the model of the Forest Stewardship Council and the Marine Stewardship Council. The Leadership in Energy and Environmental Design (LEED) is one such certified program for the building trade. But people still predominantly buy houses based on their location, their scenic view, and their amenities—not on their energy efficiency. Governments *can* nudge improvement in this regard. One interesting regulation in the UK does not permit houses to be sold unless they are brought up to a certain efficiency standard.

Many countries view carbon trading as their preferred path to reduce GHG emissions. Some countries are open to allowing trading and investments anywhere in the world that makes the most sense. Other countries such as Canada and Australia are more interested in their own domestic trading and want to keep investments within their own country. Note that both Canada and Australia have large hydrocarbon deposits of oil sands and coal, respectively, and they gain significant financial benefit from these resources. But carbon trading alone would only be effective for directing business decisions. It is not an effective means of impacting personal habits. As such, carbon trading alone will probably be insufficient to meet any mainstream targeted stabilization levels of greenhouse gases. Further, the question remains to be answered about whether governments can set an emission price high enough or a target level low enough that it will make a tangible difference in climate change without upsetting our economies or our societies. Some believe they can, while others believe they cannot.

With an expected patchwork of regulations, business will still take their best shot at what the stage will look like in the future, and they will gear their investment decisions and their relocations to match their point of view. Governments, if they are smart, should not be myopic in their thinking, and they should try to look at the world stage before enacting regulations that will negatively impact their constituents instead of simply riding a local wave of popularity. Otherwise, they are likely to do more harm than good. It is realized that effective forward thinking is difficult even for upright politicians and asking them to think ahead is asking quite a lot.

Some companies are already starting to factor CO_2 costs into future economic calculations. But even if companies are starting to play at the leading edge of regulations, they would be foolish to play at the bleeding edge and get ahead of government regulations.

California's Low Carbon Fuels Standard

One of the latest movements to reduce GHG emissions is the Low Carbon Fuels Standard (LCFS). California announced the LCFS in January 2007 to reduce the life-cycle GHG emissions of transportation fuels by 10% by the year 2020. The LCFS was approved by the California Air Resources Board (CARB) for implementation on June 15, 2007. The driving force for the proposal is that transportation accounts for 40% of California's GHG emissions, and petroleum-based fuels account for 96% of California's transportation needs. According to the original white paper report from the governor's office, the expectation is that "low carbon fuel" will replace at least 20% of the gasoline used in California.

Low carbon fuel is a shortened term used for fuels with low life-cycle GHG emissions. The hope of the LCFS is to drive fuel switching to alternative and renewable fuels, plus make upstream oil production and refining more emissions efficient. Buying and trading credits is available to those who cannot or will not be able to decrease life-cycle GHG emissions. Low GHG bio-fuels don't really exist currently, so technical innovations will be required. Several other states and three Canadian provinces have followed suit so far and agreed with the effort. The EU, Japan, Washington, and some northeastern states are watching closely to what happens in California. What is not clear is how the life-cycle emissions will be measured and by whom. In addition, how the government will actually monitor, regulate, and enforce the LCFS is still unknown. The LCFS will have the largest impact on the finished fuels resulting from oil sands crudes, coal-to-liquids fuels, Orinoco (Venezuela) extra heavy crudes, or fuels from other heavy crude oils. Unfortunately, these are the very crude oils that will become dominant in the coming years in North America and abroad.

STEVEN E. SONDERGARD

A rough comparison of options today (on a per-mile-driven basis) will be addressed. First, renewable fuels are not usually lower in life-cycle GHG emissions than normal hydrocarbon fuels. Second, ethanol from corn has roughly 10% to 20% higher life-cycle GHG emissions than normal hydrocarbon gasoline. Third, first-generation bio-diesel has higher life-cycle GHG emissions than normal hydro-carbon diesel. Fourth, normal hydrocarbon diesel currently has lower life-cycle GHG emissions than gasoline. Fifth, fuels produced from heavy crude oils and Canadian oil sands crudes currently have about 10% to 20% higher life-cycle GHG emissions than fuels produced from conventional crude oils depending on the technology used. As a subpoint, oil sand GHG emission intensity has already improved 50% since 1990. Crude oils from oil sands have a potential of having their CO_2 emissions reduced further by various prospective technologies such as in-situ SAGD, OrCrude from Opti, THAI and CAPRI by Petrobank, HTL from Ivanhoe, Vapex by Osum, and the N-Solv process. CCS and nuclear options are also being investigated by oil sands companies. Sixth, gasoline-hybrid or plug-in-hybrid vehicles are currently not as low in life-cycle GHG emissions as people might expect. This result is due in large part to the emissions from mining, smelting, and processing the exotic metals used in the batteries, the manufacturing process having double the emissions, the relatively short battery life, and the increased disposal and recycling issues. The allure of hybrids is that the actual emissions per mile driven can be about half that of conventional gasoline technology. When considering the life-cycle emissions of plug-in-hybrids, one must also consider the emissions generated from the electricity production and the transmission losses.

Overall, the individual leading efforts of California and other states can be looked at in at least two ways. These efforts could be forward thinking and lead the actions of the rest of the country and even the world. On the other hand, they are also looked at as being somewhat arrogant, and they make managing the country much more difficult, particularly if Global Warming turns out to be a less pressing issue at some point in the future relative to other societal issues. A patchwork of regulations for climate change would make it much harder to balance the impacts of the economy, energy security, and foreign policy against Global Warming.

U.S. Lieberman-Warner Climate Change Bill

This bill, sponsored by Joseph Lieberman (I-Conn.) and John Warner (R-Va.), was designed to reduce U.S. GHG emissions by 63% to 71% of 2005 levels by 2050. The bill would create a cap-and-trade system, plus limit emissions from power plants, refineries, and other heavy industry. A later amendment to the bill proposed limiting emissions from natural gas to residential and commercial buildings. It continues to be reworked, but as of this writing, if this bill is enacted as it stands, it would be the toughest global warming bill in the world and take the U.S. to the forefront of this fight. Even so, some groups are stating it is still only a starting point and more drastic reductions are needed.

The bill directs the EPA to establish the Climate Change Credit Corporation to distribute emission allowances. About 18% of these credits would be auctioned off initially, and this would gradually increase as industry adjusts. Power plants would initially be granted free emission permits, but these would be phased out by 2036. Starting in 2012, the regulated sectors would start to see their emissions capped at 2005 levels. After this date, emission limits would gradually be decreased each year until 2050. More stringent limits could also be enacted pending future scientific reviews of the system. The bill, as it stands, does not include incentives for nuclear power or renewable energy resources. Revenue streams would be available for developing low carbon or clean energy technology and helping low- and moderate-income consumers deal with rising energy prices.

The biggest fear with this or any similar bill is that it will do significant harm to the U.S. economy—which in the opinion of many is a virtual certainty. The more appropriate question is: how much harm would it cause? Some points will be described to elaborate on the magnitude. Business in the U.S. would effectively incur a new, rather large, tax burden. This tax burden has been estimated at $80 billion or more per year, or between $3 trillion and $6 trillion to 2050. Supporting politicians appear to have a keen eye on this potential pot of gold. Yet, tax policy cannot be set in a vacuum without realizing the consequences.

Businesses do not have an endless capacity to be taxed. This bill would be one of the largest income redistributions schemes ever. Some politicians don't appear to be looking at the longer-term effects. If this bill turns out to be more than a show of effort and actually passes in anything like its present form, U.S. business would be at a large disadvantage versus many foreign competitors. The future profitability of U.S. industry would be at serious risk. Costs would be passed on to consumers in the struggle for industry to survive, which will in turn negatively impact the economy and necessarily reduce the potential pot of gold. The fabric of business would be severely stressed. Businesses outside the U.S. would thrive instead. This could actually be counterproductive to the cause and increase global GHG emissions. It would discourage future investment in the U.S. Industry within the U.S. would unavoidably decline. The stocks of U.S. industrial companies would suffer relative to the rest of the world. Jobs would be lost. Retirement accounts would diminish. Inflation should be expected. Basic economic fundamentals dictate that energy prices would escalate more than they would otherwise, along with food, commodities, and the general cost of living. For these reasons, if this or a similar bill is enacted, at the very least it should have periodic openers or triggered openers to adjust its severity.

The best possible outcome for any policy requires that the probability and severity of the outcomes would be weighed in the balance against the probability and severity of Global Warming impacts, including the perceived probability of actual improvement. After an objective evaluation of available information, the weight of the answer appears to suggest it might be unwise to proceed with a bill that is as severe in nature as this one. Others have weighed the balance differently of course, and many appear unconcerned about going forward. Those in this camp must believe that the bill represents a trade-off they are willing to make or believe that the trade-offs would be largely made by others. It is also possible they may not appreciate the probable societal ramifications that were just described, or they simply desire to fulfill the present popular sentiment without regard for the consequences. It may actually be a combination of all of these.

Chapter Appraisal

Several reports, protocols, bills, and regulations have proposed aggressive reductions in human-caused greenhouse gas production. Arguments and counterarguments have both been presented herein. An assessment of the climate change problem, the probability of its resolution, plus the risk of unintended consequences, all help determine a reasonable plan. The problem of climate change does not appear to be as severe as many proclaim. The probability of successfully curtailing global warming is lower than is often forecast.

An objective evaluation suggests that vigorous action measures may have more unintended consequences and higher risks than is currently perceived. How much we should strive to reduce the impact of Global Warming must also be determined by examining the trade-offs between several societal impacts. Trade-offs between Global Warming, our economy, energy security, and our quality of life must all be made. Considering all these perspectives, a climate change plan that embraces less forceful action appears most prudent. Plans that propose further action would be very difficult to achieve and/or counterproductive either to the cause or to society as a whole.

STEVEN E. SONDERGARD

The costs of further action measures also appear to be quite excessive, particularly in relation to the benefit. After analyzing the probable upsides and probable downsides to Global Warming, based on establishing their probable risks and probable consequences, taking aggressive action is actually discouraged.

12. WHAT IS THE OUTLOOK FOR FOSSIL FUELS?

Focus: Forecasted fossil fuel reserves and production balances are presented for added perspective. A projected shortfall in crude oil supply leads to the possibility of societal stress and potential collapse. A proposed methanol economy is forwarded as one answer for a post-hydrocarbon world.

Layman's Brief:

Both natural gas and coal have known reserves that should last for more than a couple of centuries. Oil shale is also very plentiful. Crude oil, however, is another matter. We are not running out of crude oil just yet. But it is going to cost a great deal more to produce it and refine it than we have been accustomed to in the past. This is due to the expectation of more difficult technical challenges, increasing costs, and many above-ground factors.

Much of our societal benefits and synergies are the result of inexpensive fossil fuels. But fossil fuels cannot be expected to last forever. The shortfall of crude oil relative to world demand is actually projected to occur in the not-too-distant future. Crude oil is forecasted to increase in price significantly, and this will drag the price of natural gas with it. As this occurs, a huge transfer of wealth from oil- and gas-consuming countries to oil- and gas-producing countries will occur. This alone can be expected to increase societal stress to very high levels. It would be wise to prepare our societies for these pending outcomes earlier rather than later. This preparation should include research and development on post-hydrocarbon fuels. Alternative energy sources that are both economic and environmentally sound will be required for society to sustain itself.

Quick Reference:

- The global availability of natural gas is now projected to last maybe another 200+ years. Natural gas is the cleanest burning fossil fuel. Global coal deposits are relatively vast with the U.S. holding the largest deposits, and global coal is expected to last for maybe another two hundred years. But coal produces 80% more CO_2 than natural gas and 30% more than oil. Oil shale also has immense deposits, particularly in the U.S., but the technology doesn't exist quite yet for utilizing oil shale.
- Global and U.S. crude oil demand was steadily increasing until early 2008, when it started to decline. Global demand is now expected to increase only as a result of growth in developing countries. Crude oil supply by OPEC has been steady, but supply increases have been essentially limited to Saudi Arabia. Non-OPEC supply increases have also been limited. The cushion between supply capacity and demand was diminishing in early 2008, and this resulted in the corresponding price increase for crude oil.
- There is probably less than forty years to as much as eighty years before the world *effectively* runs out of crude oil—and there should be an observed global supply maximum gently projecting from a relative plateau long before that, likely within one to possibly as long as three decades. The era of

"easy oil" is nearing an end. From what is known today, a supply-demand cushion is projected to grow slightly for a few years and then diminish again in a perpetual decline, with the decline causing relatively severe global hardship.

- Assuredly before the end of this century, and possibly by mid-century, the world will need to have a radically different energy mix. To successfully navigate this fact, there will need to be multiple breakthroughs in technology in order to sustain the quality of life that much of the world has come to expect.

- The point is that nature's own fossil fuel constraints are expected to naturally limit much of the human-caused CO_2 production, which will limit any corresponding climate temperature increases. With this knowledge, it seems that what mankind should focus its efforts on are ways to conserve these limited resources and to research and develop the technologies we will need when these resources are depleted. Without these technology improvements, major adverse impacts to humans are extremely likely.

- Dollars will flow to oil-exporting countries in ever-increasing quantity and thereby transfer much of the world's wealth away from oil-importing countries. The balance of power is expected to shift in the world, and economies will be dramatically impacted. A declining oil supply is further expected to lead to famines and hardship.

- Societal collapse, one way or another, either by the disease or by the cure, appears possible. There are two ways for this not to be the case: 1) the growing popularity about the need to ruthlessly reverse man-made emissions is incorrect, *and* we do not severely harm ourselves by initiating aggressive action measures to counter Global Warming; or 2) mankind *can* control global warming and develop a less forceful plan that avoids the downsides of global warming countermeasures, and most importantly, society is successful at carrying it out and getting good results.

- Societal collapse appears essentially inevitable if we do not soon focus and embark on the R&D required for post-fossil-fuel energy technology. One method for doing this is to use methanol as an energy transport medium, for either internal combustion engines or efficient direct methanol fuel cells, or both—as Nobel laureate George Olah proposes. But energy sources would still be required—probably a combination of nuclear fission, nuclear fusion, solar, wind, hydro, and maybe others yet unknown. One advantage of this methanol fuel approach is that there would be effectively no net CO_2 emissions, because as methanol is combusted to CO_2, CO_2 could be recycled back into methanol via the energy sources.

More Detail:

The Reserve Life of Fossil Fuels

The primary focus of this book is not on fossil fuels or even economics or population growth. But these subjects are significantly correlated and interrelated with anthropogenic emissions and their impact on climate. It was realized after examining these subjects that some of their aspects needed to be covered in any balanced assessment of climate change. The analysis of climate change would not be complete without considering the impact of fossil fuels and specifically crude oils on global warming. The most probable scenario for the foreseeable future is that fossil fuels will continue to dominate our future energy supply picture for quite some time. In fact, fossil fuels are still projected to provide nearly 80% of global energy needs in 2030.[1] But how much longer can fossil fuels be expected to last?

Natural gas is the cleanest burning fossil fuel, with the lowest GHG emissions per energy unit, as it has the lowest carbon to hydrogen ratio. Natural gas currently provides about 25% of the total primary energy used in America. The proven world conventional (non-associated and associated with oil) natural gas reserves are about 6,200 to 6,400 trillion cubic feet (TCF), with world coal bed methane (CBM) reserves estimated at roughly 3,500 to 7,500 TCF. The U.S. conventional natural gas reserves are estimated at 200 to 280 TCF. The U.S. CBM reserves could be roughly 700 TCF, with maybe 100 TCF economically recoverable. In 2008 the technological breakthrough for recovering shale gas came of age. Horizontal drilling and multi-stage fracturing made it possible. Preliminary estimates of the nine largest shale gas plays in North America are now estimated to have about 260 TCF of reserve potential. Others estimate North American shale gas reserves at maybe 2,000 TCF. It is still too early to know with any certainty. Proved natural gas reserves have been increasing, particularly for North America, due in large part to coal bed methane and shale gas. The availability of natural gas is therefore projected to last another 200+ years.

Coal has a relatively high carbon to hydrogen ratio. It therefore has high GHG emissions per unit of energy, which turns out to be almost double that of natural gas. The primary demand for coal comes from the global power industry, and more electricity is produced globally from coal than any other energy source. This world demand for coal is expected to continue into the foreseeable future, particularly within China, India, and other developing countries. Coal currently provides about 24% of the total primary energy used in America. The U.S. has the largest coal reserves with almost twice the reserves of any other country, followed by Russia, China, India, Australia, and then many other countries follow with lesser reserves. The demand for coal is expected to continue, just as its domination of global power generation is expected to continue into the foreseeable future. Global coal deposits are relatively vast and expected to last for maybe another two hundred years. Another solid fossil fuel, oil shale, could possibly last even longer than gas or coal, but this depends entirely on the technology that will be eventually utilized for oil shale and the speed with which it will occur.

Crude oil is the dominant fuel source for global transportation. The supply of crude oil, however, is expected to fall short of global demand in the near future. This would create a very impacting energy shortfall. Crude oil currently provides about 40% of the total primary energy used in America. As it pertains to Global Warming, any supply shortfall would automatically limit the CO_2 emissions from crude oil derivatives. Therefore, crude oil demand, supply, and reserves will be examined in relative detail because of its importance and scale. World oil demand has been increasing ever since a small decline occurred in the 1980s. Global oil demand has been increasing at 1% to 3% per year, and it has averaged about 2% per year. In the future, world oil demand is expected to increase at 1.3% per year to 2030 (per both IEA and ExxonMobil). That would be another thirty million barrels per day by 2030 on top of the eighty-five million barrels per day currently used. This oil demand increase is expected to be primarily from developing countries, with the expected demand from industrialized countries hardly changing. In fact, 42% of the forecasted increase in demand is expected to be due to China and India. PetroChina alone expects to double its refining capacity by 2015 to 4.8 million barrels per day. In early 2008, non-OECD oil demand was remaining stable, even while OECD demand was dropping significantly.

What can halt this continued growth in global demand for oil? There appear to be at least four possibilities: 1) a global catastrophe of some kind (hopefully a low probability); 2) a severe global recession that would negatively impact the strong growth of even China and India (an increasing possibility); 3) an economic substitute for oil is developed and widely utilized (a low near-term probability); and 4) a reduction or elimination of fuel subsidies and price controls in oil-producing and developing countries that would limit their internal demand (reductions are very possible, but eliminations are a low probability). Something has got to give. A U.S. recession by itself is forecast to not be enough for a significant global oil demand reduction in the future (even though this has been the case in the past) because most of the increase in oil demand is now coming from developing countries.

Government subsidized hydrocarbon product is prevalent around the world. About 50% of the refined product produced in the world is sold at a price that is less than the price of crude oil. The governments that subsidize or set gasoline prices include China, India, Russia, Saudi Arabia, Iran, Iraq, Egypt, Nigeria, Venezuela, Indonesia, Malaysia, Taiwan, much of Asia, most of the Middle East, and most of Latin America. This practice may help stem inflation for these countries, but it decouples the natural supply and demand mechanism, which in turn magnifies the price swings of oil. All of the growth in world crude oil demand currently (2008) comes from countries that subsidize gasoline.

On the supply side, some of the major crude oil reserves in the world will be examined, as they are representative of the world picture. The world currently produces enough oil to meet the demand of about eighty-five million barrels of crude oil per day. In early 2008, there was less than a million barrels per day of extra production capacity, and it was almost entirely from Saudi Arabia. There is strong evidence to suggest the future growth in crude oil supply may be insufficient to meet future world demand. When crude oil supply cannot keep up with demand (plus a little extra for cushion), the price of oil will rise enough that demand is reduced sufficiently to meet the available supply. Cambridge Energy Research Associates (CERA) indicates that existing world oil fields are declining at 4.5% per year. The International Energy Agency (IEA) has stated that non-OPEC decline rates are 4–5% per year. Note, however, that 70% of global production comes from fields that are greater than thirty years old, and these fields are in steeper (6% to 8%) decline. Newly producing oil fields must not only make up for this supply drop but also any demand increases. The combined OPEC and non-OPEC conventional oil supply is currently struggling to grow even moderately. The energy group PIRA recently stated the conventional non-OPEC crude production has already peaked (in 2006). UBS Bank has projected that total non-OPEC oil production should peak by

2010. Current production is able to keep up with demand in large part by the production increases of light hydrocarbons called condensates, plus unconventional and very heavy crude oils. These very heavy crude oils are more expensive to produce and process.

There is common agreement that a worldwide crude production maximum *will* occur at some point in time, similar to the peak oil production which occurred in the U.S. in 1971. This U.S. peak was predicted in 1956 by Dr. M. K. Hubbert. Yet, there is a fairly wide disparity in the projections of *when* worldwide oil production levels will start to decline. Whenever a maximum does occur, a global oil production peak is expected by many observers to resemble a high plateau rather than a classic peak. The world is not running out of oil quite yet, according to CERA. But the cost of producing this oil is another matter entirely when considering the technical risks, political issues, and other above-ground factors. Some suggest (e.g. Total S.A.) oil production levels are not likely to increase much, if any, above ninety to one hundred million barrels per day.

At normal, historical growth rates, global oil demand could easily reach 100 million barrels per day by the middle of the next decade. British Petroleum contends that a peak in demand might occur before a peak in supply, but this prospect depends on poor global economics and/or significant environmental mechanisms. Indeed, as 2008 was nearing its end, world oil demand appeared to be down by an unprecedented 4%, primarily due to a global economic slowdown. The price of crude subsequently dropped so far that many crude production projects were being halted. This is dangerous, because without any supply replacement projects, a reduction in demand would need to be 4% to 6% each and every year to avoid a supply shortfall at some point.

The Energy Watch Group released an October 2007 report which stated that the world oil production peak has already occurred—although this group has a rather pessimistic viewpoint on future production rates. The date for a crude oil production maximum is often debated, but this date of maximum production is forecasted here to occur *most likely* within ten years but *perhaps* as much as thirty years from today (2008)—the later occurring with aligned changes in the economy and government policy. This projected maximum will probably not result from the depletion of the oil reserves but will be determined by above-ground factors. Even without a well-defined oil production maximum, the cushion between crude oil supply and demand is expected to be very concerning within the next few years. A shortage is therefore expected to occur sooner than we can be fully prepared for. When it does occur, crude oil will become a relatively scarce resource, and it can be expected to be priced accordingly.

The longer-term *effective* life of crude oil is projected to be between forty to maybe eighty years based on the location of remaining reserves and other factors. Unconventional crude oil sources should help push the upper limit of this range, albeit at great cost. There are actually more below-ground reserves than that, but the remaining reserves will be so concentrated by that time the politics and other above-ground factors are likely to prevent its sensible distribution, and most countries will not be able to count on constructively building their economies with it. Without an economic energy alternative, rationing oil might become quite common. Energy conservation and rationing might be required simply for the extended economic survival of many, and it can be expected that the world situation during this time would become quite tense. So tense, in fact, that it would be surprising if wars did not break out over oil at some point during the twenty-first century. Some believe that the current war in Iraq is a forerunner of this and that the tensions from this war could escalate. Obviously, a nuclear war could make any arguments about climate change moot.

An examination of the largest oil-producing fields and countries is believed to be representative of the world oil situation, and these are discussed below:

- Saudi Arabia Ghawar—this is the largest conventional oil field in the world. It was discovered in 1948, and it started producing in 1951. It is estimated to have about 80 billion barrels of reserves, but the country keeps its official reserve amounts a secret. Many geologists having familiarity of the Ghawar field indicate it has reached a plateau and speculate this field should be in decline within the next five years. The field is currently in water flood, and it is reported to have between 25% to 35% water cut. The Ghawar field produces only about half of the oil produced in Saudi Arabia. The country is estimated to have the largest conventional reserves in the world, about 260 billion barrels, and it has already produced about 100 billion barrels. Saudi Arabia currently produces about 9.7 million barrels per day, with suspected capacity at just under 11 million barrels per day and with plans to get to 12.5 million barrels per day by about 2010. The hope is that Saudi Arabia can increase its production to 15 million barrels per day by 2020, with further increases later, as this country has the vast majority of the world's spare capacity. But Saudi Arabia has indicated that 12.5 million barrels per day is all they are planning for at this point.
- Kuwait Burgan—this is the second largest conventional oil field in the world. It was discovered in 1938, and it started producing in 1948. The field has 55 billion barrels of proved reserves and produces at about 1.7 million barrels per day. The country has reserves of about 97 billion barrels and produces at about 2.5 million barrels per day. Kuwait has 124 discovered fields including deep water offshore. But eight major fields account for about 90% of its production. Enhanced Oil Recovery (EOR) and heavy oil production have yet to be pursued. The country desires to invest some $63 billion in its fields and infrastructure in order to bring its production rate to three million barrels per day by 2010. A shortage of people is cited as their most critical concern.
- Mexico Canterell—this is the third largest conventional

oil field in the world. This field currently produces less than 1 million barrels per day, down from a high of 2.4 million barrels per day in 2004. It is in tertiary recovery (nitrogen injection), and it is now in very steep (15%/yr to 20+%/yr) decline. It has an estimated remaining life of less than a dozen years. Total Mexican crude production was about 3.0 million barrels per day in 2007, with exports at about 1.8 million barrels per day in 2006 and 2007 at about 1.5 million barrels per day. Mexican crude exports are expected to decline to less than 1.4 million barrels per day in 2008. Without capital expenditures for Mexican crude oil fields, Mexican crude production could fall by as much as one million barrels per day within a decade. With significant expenditures, it could remain roughly flat for a decade, but the probability of these expenditures being made is not considered high.

- Venezuela Orinoco—this is arguably the largest unconventional oil field in the world, with approximately 2,500 billion barrels of estimated resource and unofficially 200–400 billion barrels of probable reserves, depending on recovery ratios (The Orinoco doesn't yet have certified proved reserves.). The region has a potential life of a few hundred years, but this depends on its production rate, and that is highly dependent on Venezuelan political stability and other economic drivers. This region produces oil that is about 9°API gravity (extra heavy oil), and it must be diluted just to be transported in a pipeline to the upgraders located at José. The Orinoco is only one of several oil regions in Venezuela. The Lake Maracaibo region has significant but declining production—crudes from this area are blended to gravity and are termed the Bolivarian Coastal Fields (BCF). The country produced roughly 3.3 million barrels per day in 2002 prior to the countrywide oil strike, but it has been steadily declining and is now reportedly closer to 2.5 million barrels per day. The political climate in this country is now viewed as too unstable for much International Oil Company (IOC) participation. The combined property nationalization, taxes, and royalty policies are expected to limit future crude production. With very large estimated reserves, Venezuela has big plans for crude production increases, but its political stance makes it difficult to project the likelihood of this occurrence or its timing.

- Canada Athabasca Oil Sands—this is the second largest unconventional oil field in the world with an estimated 1,700 billion barrels of total resource and an official 173 billion barrels of reported reserves. This region produces a very heavy bitumen (about 8° to 9° API gravity) material that must be upgraded, similar to the unconventional Venezuelan Orinoco reserves, before the material can be processed as conventional crude in most refineries. This basin started initial production in 1967, and it was not economic for several decades. It has experienced rapidly increasing production as a result of the recent run-up in crude oil prices plus the diminishing drilling opportunities elsewhere. The Alberta oil sands currently produce over one million barrels per day, with projections to reach two million barrels per day by ~2010, and potentially three million barrels per day by 2020–2030. The expected life of this deposit is eighty to one hundred years at projected production rates, but it currently appears production rates will be limited more by economics, water availability, and air emissions.

- Russia—the largest non-OPEC producer of crude oil with production at roughly 9.8 million barrels per day. This rivals Saudi Arabia as the largest crude producer. Crude exports from Russia are now over four million barrels per day, with another two million barrels per day of product exports. Russia is an increasingly important supplier to Europe, supplying almost 30% of Europe's crude slate. However, Russia's crude exports are expected to start a decline simply due to increased internal refinery requirements. USSR crude production declined rapidly after the iron curtain fell, from over eleven million barrels per day down to less than six million barrels a day in the mid 1990s. For the last decade, Russian crude production has been increasing by between 3% and 8% per year, but this growth is now slowing, and some analysts expect Russia may soon begin a crude production decline. This decline in crude production is expected due to recent Russian policy factors, including the Yukos situation. The uncertainty of Russian taxation as well as the limitation on investment capital are cited as other causes.

- North Sea—this region includes areas controlled by the UK, Norway, Denmark, Germany, and the Netherlands. Oil was first discovered in this region in the early 1960s and started producing in 1971. Production picked up greatly in the 1980s. Drilling in this region was extremely hazardous due to weather and sea conditions. Brent crude oil comes from this area, and it is now a world benchmark crude and used in many crude pricing formulas. This area had peak production in 1999 at about six million barrels per day, and it has been in decline since that time, with more recent declines at about 10% per year.

- U.S.—The U.S. is the third largest producer of crude oil in the world, producing over eight million barrels per day, behind only Saudi Arabia and Russia. However, the U.S. has such an appetite for oil, approaching 21 million barrels per day, that it imports even more oil than it produces, over 12 million barrels per day of imports in 2007. The crude production from the U.S. has been in decline since 1971, and this decline is expected to continue. Current policy and public sentiment are contributing factors. Drilling is not always allowed where the prospects of finding oil are the most probable or the least costly. Roughly 85% of the lower-48 U.S. offshore oil production prospects are off limits to drilling. Other large albeit sensitive areas (e.g. ANWR) are also off limits to drilling. The U.S. is the only crude-importing country in the world that has restricted access to potential reserves, limiting their domestic crude production. Further, these restricted potential reserves are currently believed to be quite large. Perhaps the U.S. is still too rich and oil prices need to rise even more before the priorities change.

- Prudoe Bay—this is one of the largest U.S. conventional oil fields, producing Alaska North Slope crude (ANS). This field was discovered in 1976, and it required the construction of the Trans-Alaska Pipeline System (TAPS) to Valdez before full production could be achieved. The estimated reserves from this field are about 13 billion barrels. The field produced at almost 2.1 million barrels per day in 1988, but it now produces less than 800,000 barrels per day, and production is declining at over 6% per year.
- West Texas Intermediate (WTI)—this region has multiple fields, and many started production in the 1930s. WTI crude is a light (~36°API), sweet (0.6% sulfur) crude oil. This region has been significant enough that it is one of the dominant world benchmark crudes, and it is used as the benchmark crude on the New York Mercantile Exchange (NYMEX). The West Texas region has been in crude oil decline for decades.
- Gulf of Mexico (GOM)—this region has relatively large reserves and produces about 1.3 to 1.5 million barrels a day. Offshore drilling in the gulf continues into deeper and deeper water as the technology to do this deeper drilling improves, albeit at greater cost. Production levels from this region should increase for several years to come, but the Gulf of Mexico region is projected to start a decline within a decade.

- U.S. oil shale—the world's largest oil shale deposits are in the Colorado/Utah/Wyoming area. This deposit is actually quite large, with an estimated 1.5 trillion barrels of oil in place and 800 billion barrels of recoverable reserves, per Argonne National Laboratory. At a production rate of four million barrels a day, which is a rough projected limit due to electricity and water availability, these reserves could have a life of several hundred years. But recovering this resource is not yet technically or economically viable.

STEVEN E. SONDERGARD

The U.S. Department of Energy (DOE) has helped fund several potential processes with the Federal Research and Development Program. Shell Oil has recently pioneered a promising new approach of in-situ retorting, whereby electrical resistance heaters are lowered into drill holes in the shale bed in order to heat the shale to about 600°F and convert the raw kerogen, an oil precursor, into useful hydrocarbon. The Shell process is projected to perform this conversion over a period of four to five years, versus millions of years when left only to nature. Small-scale testing of this process does show promise, and it is now in extended larger-scale tests. The results of these tests should provide answers in about five years on whether the technique will work at scale. Other oil shale demonstration efforts are underway by Chevron and the American Shale Oil Corp. If any of these processes work well, there are oil shale deposits in more than twenty other countries that could benefit.

Figure 12.1 is an illustration of the countries and regions that have already experienced their conventional crude oil production maximum. Figure 12.2 graphically depicts the production levels of major international oil companies and how flat their production levels really are— and this is with the top oil companies reinvesting their recent record profits back into further exploration and production. As unpopular as oil-company profits are to many people, any reduction in profitability will necessitate a reduction in future exploration and production.

It should be apparent from this data that the future supply of crude oil cannot be expected to increase exponentially, and the supply of crude oil will be hard-pressed just to meet world demand. Furthermore, an exponential increase, or much of an increase at all, in CO_2 emissions from the combustion of crude oil derivatives should not be expected either.

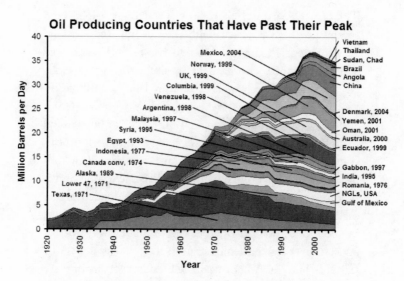

Figure 12.1. Oil-producing countries that have passed their maximum production. (Adapted from: Report to the Energy Watch Group, October 2007, EWG-Series no 3/2007, p.11)

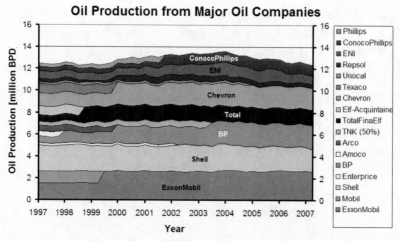

Figure 12.2. Note that the production levels of major oil companies have remained relatively flat, with a gentle peak in 2004. (Adapted from: Report to the Energy Watch Group, October 2007, EWG-Series no 3/2007, p.11)

STEVEN E. SONDERGARD

Contrary to popular belief, oil prices are not set by major oil companies. Major oil companies do not control the price of oil or natural gas any more than you control the price of your house. The factors that set the price of oil are complex, but the actual supply/demand balance is a primary factor. The near-term supply/demand balance is known as *the fundamentals*. The *perceived* or *projected* supply/demand balance is also a strong influencing factor on price. This projected impact on the fundamentals can be influenced by many above-ground aspects. These above-ground aspects include government policy, climate policy, monetary policy, exchange rates, steeply rising costs, capital and credit issues, investment patterns, environmental issues, geopolitics, strikes, labor shortages, wars and skirmishes, increasing technical requirements, depletion policies, supply chain infrastructure issues, and severe weather. There is an estimated 1.2 trillion barrels of reserves still in the ground, but above-ground factors are much more uncertain and increasingly more important than below-ground reserves.

Oil is becoming increasingly more expensive to find and produce. The era of "easy oil" is coming to an end. In particular, in order to meet world demand, production will need to be developed from oil reservoirs that have structurally higher production costs. Drilling costs have gone up 300% to 400% since 2004. The cost of producing incremental upgraded oil sands crude has gone up by a factor of three to four in the last decade. Gulf of Mexico production costs have gone up similarly. The price of oil must rise to the point that these incremental barrels become economic, or they will not be developed. A floor price of oil is therefore created by the most expensive barrels needed to be produced in order to meet the world demand. The price of oil can be elevated further above this floor by the perceived impacts of the above-ground aspects and commodity speculation. The price of oil can drop below this floor only for relatively short periods. Delays in oil production investments can result from signals of uncertainty in either demand or price, and it should be noted that investments in increased production will be limited unless the short- and long-term price is viewed as sustainable enough to recoup these investments.

OPEC's National Oil Companies (NOCs) currently hold 77% of world crude reserves, and non-OPEC NOCs control another 8%. National Oil Companies will increasingly control even more of the world's oil reserves in the future. OPEC countries and NOCs will therefore have increased *economic* control as time marches on. With a little knowledge, the ire of the public might transfer from "big oil" to NOCs.

It can reasonably be expected that some people within oil-producing countries and regions will increasingly try to hoard these resources for internal use or try to increasingly control the economics of their resources. Alas, this is an outcome of human nature—simple greed and fear. We have already seen the initial stages of these actions. Countries, NOCs, and the citizens of regions having economic hydrocarbon reserves are already more hesitant to share their bounty with others outside their region. This translates into increased royalties, increased taxes, the ejection of International Oil Companies (IOCs), and additional regulations, royalties, or taxes to dissuade economic value from leaving a country or province. Dollars will flow to oil-exporting countries in ever-increasing quantity and thereby transfer much of the world's wealth away from oil-importing countries. The balance of power is expected to shift in the world, and many OECD economies will be dramatically impacted.

Crude production levels are further inhibited when the most knowledgeable people and companies are excluded from participating or helping in the efficient development of the resources within some countries. NOCs often have political drivers other than economics, and NOCs are (in general) about two-thirds as efficient at exploration and production as IOCs. The technical and commercial competence of NOCs is widely variable, but it *has* been increasing. The international focus of NOCs is also expanding. NOCs increasingly prefer to deal with other NOCs, and IOCs are being allowed access to less and less of the world crude oil reserves.

Relying on OPEC to increase world production or threatening them with some sort of penalty shows naïveté about the situation— in simple terms, OPEC countries are for the most part at cultural, political, economic, or technical limits, and they essentially *can't* increase their production. For this reason, energy policy will need to amount to much more than pleading with OPEC or even just Saudi Arabia to increase supply.

In the first half of 2008, the crude oil supply-demand balance was relatively tight, causing the crude oil price to escalate. It is sus-pected that some of this price increase was due to speculative trad-ing, but that does not diminish the underlying shortage. Crude oil prices started to fall when the supply-demand cushion expanded. The subsequent price decline correlated with falling world crude demand. In the near-term future, downward pressure on the price of crude oil could last a few years, but this duration will depend on both the severity of any economic downturn versus the supply-demand imbalance, and on any OPEC (and non-OPEC) price sup-port mechanism to reduce crude oil supply. OPEC countries could *conceivably* cooperate to effectively reduce the crude oil supply-demand cushion in order to support a price of crude oil at virtually any value. But OPEC restraint is unlikely for financial and geologi-cal reasons, and high compliance with a significant reduction target would be unprecedented (in other words, their incentive to cheat is high). After a short-term "glut," the escalation of crude prices should again occur. Within less than a decade, the world crude oil price is expected to again increase to unprecedented levels as a result of a restricted crude supply compared to demand. This projected shortfall is expected to cause relatively high economic stress on a global scale—at least until technological breakthroughs can provide a reasonable energy alternative to crude oil.

The evidence should be clear about fossil fuels. Assuredly before the end of this century, and reasonably before mid-century, the world will need to have a radically different energy mix. Successful navigation of our future will need to include multiple breakthroughs in technology in order to sustain the quality of life that much of the world has come to expect. To start with, mankind should strongly focus its efforts on ways to conserve our limited fossil fuel resources, particularly the resources with lower life-cycle GHG emissions. North Americans in particular will need to be much more energy efficient in their use of crude oil and natural gas. In parallel with fossil fuel energy conservation, new technologies will need to be developed before these resources are depleted. Without these technology improvements, major adverse human impacts are extremely likely. A carbon tax or a cap-and-trade system may indeed be effective in altering the balance between economics and the environment and move the balance toward achieving increased energy conservation. But great care should be taken in the application of these policies in order not to negatively impact the world's economies or diminish a country's competitive position relative to the rest of the world!

Many in the environmental community have taken the idealistic stance of restricting domestic fossil fuel production. It should now be apparent that this action would classify as a dangerous high-risk measure. The probable impacts of reduced domestic crude oil production have a much higher downside than upside for North American society. It should be realized that retaining economic prosperity in general is at least as large a societal driving force as are environmental concerns. That alone may be hard for some to accept. A probable supply-demand shortfall in world crude oil can be expected in the not-too-distant future. Even without any decline in U.S. crude oil production, a forecasted restriction in world crude supply has a high probability of negatively impacting the U.S. economy, due to the fact that the U.S. is currently such a large net importer of crude oil. If cap and trade—or any other fossil fuel tax—is enacted, it is all the more important to support domestic production to help feed the goose laying the golden eggs. (It is important to realize that only business creates basic wealth in a free market. Food, energy, and shelter are the foundational businesses. Food is being threatened and subsidized, shelter is stagnating, and that leaves energy.) If this goose is killed or gets sick, the economic engine and revenue-raising mechanism could cease prior to an alternative goose being developed to lay, and our goose would be cooked.

Restricting domestic drilling is dangerous, even with an environmental-only perspective. When economic prosperity is threatened with reduced domestic production of oil and gas, a probable scenario is that society could be driven to increase the combustion of other fossil fuels that have even higher CO_2 emissions. Increased greenhouse gases would result if reduced domestic conventional oil supplies promoted more utilization of coal, heavy oils, or possibly oil shale, without effective, efficient, and inexpensive carbon capture and storage. This would employ fuels with much higher life-cycle GHG emissions, and capturing their increased CO_2 emissions would be secondary. Further, the environmental risk of oil spills from new off-shore oil wells, even wells drilled in deep water, has been virtually reduced to zero. For instance, every oil shutoff valve worked perfectly during the devastating hurricanes Katrina and Rita in the Gulf of Mexico. Domestic drilling therefore reduces oil imports, improves energy security, reduces the trade deficit, boosts U.S. revenue, helps the economy, increases jobs, has a net environmental benefit, and bridges us to alternative energies.

Greenhouse gas emissions can be directly traced to both the combustion of fossil fuels and our globe's large population. These factors are further linked because the world's high population today is only sustained because of the existence of large quantities of inexpensive energy. Affordable energy provides vast improvements to the quality of human life. The utilization of Earth's fossil fuels has increased our ability to produce food, recover and purify water, provide shelter, and allow specialization of human tasks. A declining oil supply will by necessity put great pressure on the planet's ability to sustain our current, much less expanding, population. Projected future population increases will only make the impact of expensive energy more severe when it occurs.

The development and production costs for new oil reserves have also been increasing rapidly and causing the price of oil to increase along with it. This simply means that even while oil is still available it could get much more expensive. Without affordable energy, many people would face significantly greater hardships—in essence, life for the majority of humans would be very tough and even brief. People who cannot afford energy would be the ones to suffer, and the poorest would suffer more. Food and water would become relatively scarce as well, and again the poorest of the population would be impacted most. A declining oil supply is further expected to lead to severe famines and universal hardship. This might be expected to cause the massive loss of lives, perhaps ultimately the loss of the majority of humans.

To summarize fossil fuels, natural gas and coal both have large domestic reserves to supply America's energy needs through this century. The near-term supply of crude oil, however, is a different story. Domestic drilling may be the most viable mechanism to circumvent a shortfall prior to economic alternative energies becoming available. A near-term crude oil shortfall would surely create economic hardship. But, as it relates to Global Warming, any shortfall would limit the CO_2 emissions from crude oil derivatives. Further, when economies are wobbling and people groups are threatened, the basic survival instincts will become elevated in importance for many people, and they could override all other societal priorities. Concern for our globe and the environment would consequently be reduced in significance. The interrelationships between economics, energy, population, and climate change can therefore not be ignored. The points mentioned here might seem dramatic, but they are intended to highlight some of the interrelationships and the delicate balance of opposing factors through which the best policies must navigate in order to avoid severe consequences.

A Future Methanol Economy

After crude oil is gone, and even before it is gone, our societies will need to be based on some other form of energy and transport. George Olah et al. describe one such system in their book, *Beyond Oil and Gas: The Methanol Economy*, 2006.[2] The full vision of a methanol economy includes several facets. First is the production and utilization of methanol and/or dimethyl ether (DME). Second is utilizing exhaust CO_2 from combustion processes to remake methanol or dimethyl ether again. Third is transporting and using methanol and dimethyl ether as transportation fuels. Fourth is using methanol as a raw material for ethylene, propylene, and other petrochemical feedstocks.

Methanol is proposed as an energy transport medium—for either internal combustion engines, proton exchange membrane (PEM) fuel cells, or efficient direct methanol fuel cells (DMFC), or possibly all of the above. Energy sources would still be required—probably a combination of nuclear fission, nuclear fusion, solar, wind, hydro, and maybe others yet unknown. One advantage of this methanol fuel approach is that there would be effectively no net CO_2 emissions, because as methanol is combusted to CO_2, CO_2 would be recycled back into methanol via the energy sources.

Using methanol as a fuel would actually have several potential avenues. Methanol is an excellent transportation fuel in internal combustion engines. It has been used in Indy race cars and dragsters for decades. It has a higher flash point, and methanol fires are easily extinguished with water. Methanol is a convenient medium to store energy, and it can be transported with relative ease. Liquid methanol could also be directly fed to a fuel cell without first converting it to hydrogen. Currently, methanol is the only liquid fuel than has been demonstrated to have practicality in fuel cells for transportation applications. The technology for DMFC has already been developed in a cooperative effort between the Caltech-Jet Propulsion Laboratory of NASA and the University of Southern California.

DMFC technology will assuredly get even better in the future as research proceeds. More recently the process was found to be reversible, which may become important in making the methanol.

Fuel cell vehicles promise to be quieter, cleaner, and require less maintenance than conventional internal combustion engines because they have fewer moving parts. PEM fuel cells operate on hydrogen as a fuel and are currently favored because of their lower weight, low operating temperature, and high power output. But PEMs require either upstream reformers to convert a liquid fuel to hydrogen or they must transport and store hydrogen directly. Reforming or storing hydrogen not only adds weight but increases safety concerns. If PEMs are used, methanol is the most promising liquid fuel to be stored in vehicles and then converted to hydrogen on board. However, at this time, hydrogen from methanol reforming contains over 100 ppm of CO, which poisons PEM fuel cell catalysts operating below 210°F, and the CO must be removed prior to injection into the PEM.

Onboard reformers are currently about 80% efficient, but they have their own challenges. They require high temperatures to operate properly, and they need considerable time to heat up and reach these temperatures. Methanol steam reformers operate at lower temperatures (480°F–660°F), but they are still relatively expensive. Work is progressing on reducing the CO levels in reforming methanol to hydrogen. Autothermal reforming of methanol, which combines steam reforming and partial oxidation, also holds some promise.

DMFCs get around many of these common problems with PEMs. Several prototypes of DMFCs have been constructed and tested to date. The DMFC consists of two electrodes separated by a PEM and connected to an electrical circuit. The anode contacts the stored methanol-water mixture, where it is oxidized to produce protons that travel through the PEM by ionic conduction. The electrons then travel through an electrical circuit. The cathode contains a platinum catalyst that is in contact with air or oxygen. The disadvantages of DMFCs at this point include their weight, complexity, stacking issues, and cost—but there is much promise for the future. Their advantages include excellent reliability, long service life, and minimal maintenance. Improvements are being made in both cost and performance. General Electric (GE) and others are also working on DMFCs for home and small appliance power applications.

In addition to being directly used as a fuel, methanol can be easily converted to dimethyl ether, which has a relatively high energy content and a high cetane value of 55, making it an excellent substitute for diesel fuel.

There are several promising new technologies that could capture CO_2 from the atmosphere and convert it into methanol. One such process could utilize algae or other microbes to metabolize CO_2 and H_2 to form methane, which could then be converted to methanol. Genomics to modify these microorganisms could make the process more efficient. There are other processes that incorporate photosynthesis that could also be used.

Once the issues with the utilization and production of methanol from captured CO_2 are conquered, the transportation of the methanol and DME would be relatively straightforward, as existing pipelines, elastomers, and technologies could be utilized. In addition, the production of chemicals from the methanol, or even methane, would also be a relatively simple extension of current refining processes to then achieve the full vision of a methanol economy.

13. WHAT CAN WE REASONABLY DO?

Focus: Specific action measures are presented. The Royal Dutch Shell Trilemma is relayed. The Vattenfall perspective is discussed. Economic measures and energy efficient measures are mentioned. Vehicle fuel efficiency and other transportation sector improvements are analyzed, including hybrids. Nuclear, coal with carbon sequestration, coal-to-liquids, and hydrogen are considered. Renewable fuels are covered. Ethanol (including cellulosic), bio-diesel, Jatropha, bio-butanol, algae, wind, solar, and waves each have mention. Finally, simple estimates are formulated to project future CO_2-equivalent emissions for both the world and the U.S.

Layman's Brief:

Increasing population and forecasts of increasing economic prosperity set the stage. Without both of these assumptions, the problem is necessarily reduced. In order to decrease atmospheric GHG levels in this century, the world will have the challenge to supply almost twice the energy needs at roughly half the GHG emissions. Many then argue that it is better to be safe than sorry, and they desire to err on the side of overreaction. Others argue the opposite, stating in effect that we should first understand what we are jumping into. Yet, there are some action measures that probably make sense whatever one's stance. This chapter mentions many of the most-prescribed action measures. It is also relayed that single solutions won't get very far. A portfolio of action measures will be required. Both developing countries and industrialized countries will need to have a part, and neither individuals nor industry will be exempt from the pain. Research and development should be pursued on multiple fronts.

A balanced and realistic view of global warming suggests it may be safer to under-react than to overreact. This is because it appears anthropogenic Global Warming may not be as severe as is often portrayed, that we may actually have a higher GHG threshold than most prescribe, and that the economic and societal impacts of many mitigating action measures may be quite severe. These points lead us to embrace a plan of no-harm action and low-impact action as most prudent and politically achievable, worldwide. Yet, even these less-extreme actions still allow for quite significant improvements in the emissions of anthropogenic greenhouse gases and for delaying the negative impacts of forecasted outcomes.

Quick Reference:

- No-harm and low-impact actions could easily include better insulation, more efficient lighting, improved air-conditioning and heating, more efficient water heating, improved vehicle fuel efficiencies, nuclear power, sensible carbon sequestration, better community planning, waste minimization, avoiding deforestation, and changing wasteful lifestyles.
- Adding insulation in existing buildings and requiring better insulation in new construction are the most cost-effective measures to reduce GHG emissions. Indeed, these measures are quite economic, and they would increase GDP if implemented.
- The ongoing improvements in energy efficient lighting are

particularly dramatic. Fluorescent lighting is more efficient than incandescent lighting. New light emitting diode (LED) technology is even better yet—already being multiple times more energy efficient per lumen than incandescent lighting. Improving lighting may sound small, but 19% of all generated electricity is used for lighting purposes.

- Better community planning would allow less driving and more closely knit communities if businesses and services were closer to homes. Less transportation fuel would be used, and it would be less wasteful in general. Communities could also require or encourage the use of additional insulation, resulting in less wasted energy for heating and cooling.

- Industry in developed countries is already on a path of improved energy efficiency and reduced emissions, primarily due to the economics of increased fuel costs but also due to the environmental concerns and driving forces.

- Many parallel action measures must be successfully pursued in order to have a sizeable dent in emissions. There are not any action measures that are large enough by themselves to have more than minimal impact. In fact, the 27 giga tonnes of known and reasonable action measures will only be a small portion of the most-prescribed reduction requirement.

- Selectively targeted policy for various action measures appears to be the best approach. Cap and trade or a carbon tax will not be effective across the entire spectrum of action measures required. It turns out the pain of action measures will need to be shared (about equally) between individuals, industry, and government.

- About 10% to 20% of the emissions from hydrocarbon fuels come from the production, manufacture, and transportation of the fuels, while about 80% to 90% of the emissions come from the *use* of the fuel. In addition, the average North American automobile weighs nearly 4,000 lbs, while the average North American weighs less than 200 lbs, and most vehicles carry only one person most of the time. Therefore, most of the emissions from transportation fuels are a result of the work required to propel the vehicle rather than for propelling the occupant(s). Reducing the average weight plus reducing the wind resistance of automobiles could substantially improve the emissions per occupant-mile driven. The technology to achieve this already exists. Performance does not need to be sacrificed. Many European cars and cars sold in most countries outside North America are smaller and lighter and have better mileage relative to the average North American car. In support of this idea, the National Petroleum Council (NPC) unveiled a report in July 2007 declaring that CAFE standards could double in the U.S. by 2030.

- Ethanol fuel is touted as a fuel of the future because it is renewable. The University of California (Berkley) reviewed the literature and concluded recently that a gallon of ethanol is about 10% to 15% better for GHG emissions than a gallon of normal hydrocarbon motor gasoline. However, it must also be understood that ethanol has only about two-thirds the energy content per gallon compared to normal hydrocarbon motor gasoline. Therefore, based on the available energy content, ethanol has about 10% to 20% higher GHG emissions than normal hydrocarbon motor gasoline.

- Ethanol from corn is not very energy efficient. It takes about 75%-85% of the energy to make corn-based ethanol than is available in the ethanol itself (assuming a modern ethanol plant that uses molecular sieves to separate the water from the ethanol). For comparison, ethanol from Brazilian sugarcane requires only about 14%-20% of the energy input that is available in the ethanol. The main reasons for this four- to sixfold difference include: the sugar content is higher in sugarcane versus corn, Brazilian sugarcane can get multiple harvests per year (up to three harvests per year), sugarcane does not need planting every year like corn (the last two points decrease the plowing and planting energy/ emissions versus corn), and the waste material from the sugarcane is burned in the local Brazilian communities as fuel.

- In addition, ethanol from corn is a trade-off with the

food supply, as about 24% of the U.S. corn crop went to ethanol production in 2007. In July 2007, there were 119 operating ethanol plants in the U.S., with eighty-six under construction, to yield a total of 6.4 billion gallons per year of ethanol—just short of the 7.5 billion gallons per year by 2012 mandate from the Renewable Fuels Act. Proposals for increasing ethanol production further (by roughly five times and equivalent to about 100% of our corn crop) were finalized by the Renewable Fuel Standard (RFS) in late 2007—but with what impact on food prices?

• There are high hopes to soon be able to economically make so-called cellulosic ethanol from ligno-cellulosic materials (newspapers, wood chips, switch grass, etc.). There is significant effort currently to overcome the predominant problem of breaking down and converting the hemi-cellulose molecules to sugars. There are several known ways this conversion process can be done, including strong acid, weak acid, gasification, and enzyme techniques. Each of these processes is technologically feasible today, but none of these processes are very efficient, either thermodynamically, biologically, or economically.

• Bio-diesel is held up as an answer. This material is predominantly produced from waste animal fats to yield fatty acid methyl esters (FAME) in the diesel boiling range. However, this fuel also has its drawbacks. There are limited feedstocks available, it requires government subsidization to be economic, it has a poor cloud point and pour point (it doesn't flow well in cold weather), and it must be blended into normal diesel at less than about 5% in order to still be able to meet American Society for Testing and Materials (ASTM) diesel fuel specifications. In July 2007 there were ninety-six plants scheduled to be online within eighteen months in the U.S. to yield a total of 3.28 billion gallons per year.

• The revival of nuclear plants appears to be imminent. There are currently 104 commercial nuclear reactors operating in the U.S. (440 worldwide). However, the U.S. Nuclear Regulatory Commission (NRC) has recently received its first permit application since the 1970s and since Three Mile Island in 1979. The NRC has received seventeen combined operating license applications, which represent twenty-six nuclear units. This new wave of new reactors will be considered within the next three years or so. But it will take *at least* ten to twelve years before any of these plants could be started up, due to the time it takes to permit and to construct a plant, and this assumes no extended opposition.

- Carbon sequestration—otherwise known as carbon capture and storage (CCS)—has been proposed to reduce the net CO_2 emissions from stationary sources, of which power generation is the largest contributor. Evaluations of CCS by the USDOE/NETL indicate the U.S. has about six hundred years of natural CO_2 sinks, and the world has thousands of years of CO_2 sinks. These include deep geologic saline formations, depleted oil and gas fields, and coal seams. Coal currently produces 50% of America's power, 70% of India's power, and 80% of China's power. The NETL projects that, by 2030, coal will dominate the CO_2 production from power generation at 88%, natural gas will account for 10% of the CO_2 from power generation, and oil will account for 2%. Thus, sequestering CO_2 from coal-fired power plants could have a huge CO_2 reduction benefit. Yet, with current technology, the energy requirements to operate the CCS plants could be from 14% to an astounding 40% increase in energy needed relative to a pulverized coal (PC) plant without CCS—just to capture its flue gas CO_2. In addition, a 2007 study at MIT suggests electricity generated from PC power plants with CCS would be 60% more expensive than from PC plants without CCS. Even electricity generated from a coal-fired integrated gasification combined cycle (IGCC) plant with CCS (clean coal) would be 35% more costly than without CCS. Significant breakthroughs in technology will therefore be required for coal-fired CCS to work well. In addition, the risks and environmental impacts of CCS should be carefully assessed because Earthquakes or volcanic activity could suddenly release the CO_2 from underground formations.

- Coal-to-liquids (CTL) is another option, where coal is

STEVEN E. SONDERGARD

converted to a synthetic gas and then recombined into liquid fuels (gasoline, jet fuel, and diesel) via the proven Fischer-Tropsch (FT) process. This is a relatively capital intensive process (with promoters seeking government subsidization), and it yields quite high CO_2 emissions per barrel without CCS. The process produces very high cetane diesel, good quality jet fuel, and has a potential to reduce imported crude oil.

- Hydrogen is widely touted as a future fuel. Pure hydrogen indeed burns cleanly to water vapor, and it is particularly efficient when fed to proton exchange membrane (PEM) fuel cells. But hydrogen is not without major problems. Hydrogen is highly combustible, and it is dangerous to transport and store. It would require a completely new supply system. Hydrogen is most economically obtained from natural gas (methane) via steam methane reformers (SMR). Hydrogen can also be obtained by electrolyzing water, but producing hydrogen in this manner would use more energy to produce than is available in the hydrogen itself, which results in the hydrogen effectively becoming a simple energy transport medium. Methanol (or even methane) would have better properties and potentially existing distribution systems for that.

- A realistic improvement in CO_2 emissions resulting from multiple action measures appears to be in the order of 15% to 30% for the U.S., with a 50% to 90% global *increase* in CO_2 emissions by 2050. This is far short of what is prescribed by many. The reason for the shortfall is due to projected GDP and population increases that overwhelm achieved improvements.

- What we should not forget is that there are always trade-offs in the consequences of everything we do, and we should try to understand these consequences and quantify them as best we can before we initiate action. It is unfortunately the case far too often that we are negatively impacted by an unforeseen consequence to a popular or legislated action. This is known as the Law of Unintended Consequences.

More Detail:

Multiple, radical, game-changing technology improvements will be required to meet targeted levels of CO_2. Although perceived as unlikely by some, research and development on these multiple technology improvements should be pursued with diligence and be supported by government policies to allow the best chance of success. Some second-generation bio-fuels including algae grown for fuel show particular promise. Next generation nuclear energy also appears to hold much potential and there is high hope for cellulosic ethanol.

There are some very simple action measures that should help in the meantime. Energy efficiency measures should become a prime focus. These measures include insulation in buildings, increased transportation efficiencies, improved driving habits, improved machinery efficiencies, more energy efficient appliances, and conserving fossil fuels. Other measures that are more complex include improving power generation from coal, gasification, utilizing carbon capture and storage, and incorporating next-generation renewable energy sources. A good argument could be made that every one of these measures will need to be pursued with vigor. Each of these measures, and more, are discussed below.

The Trilemma

A concise way to think about the conundrum of Global Warming is forwarded by Royal Dutch Shell Oil Company.[1] The world population is expected to approach nine to ten billion people by 2050. In a nutshell, the world will have the challenge to supply almost twice the energy needs at roughly half the GHG emissions. That translates into a requirement for delivering energy with one-fourth of the CO_2 emissions per unit of energy. That will be a truly enormous task.

Shell describes the Global Warming problem as a "trilemma." Any solution will have varying degrees of the contrasting characteristics of being clean, cheap, and convenient. But there isn't a solution that perfectly meets each of these characteristics, and every known solution gives up one or more of these, at least to some degree. Between *convenient* and *clean* there are alternative technologies such as solar, wind, bio-fuels, nuclear, and waves. These technologies themselves inherently provide the benefit of reduced GHG emissions; however, they are not yet the most economic alternatives. Between *cheap* and *convenient* there are alternatives such as coal, heavy crude oil, and oil sands. The mitigating measure for these alternatives includes carbon capture and storage in order to reduce their GHG emissions. Between *clean* and *cheap* there are efficiency improvements such as increased insulation, improved vehicle efficiencies, improved lighting, improved space and water heating, and improved air-conditioning. At various locations within the triangle there are alternatives such as hydro and geothermal, which are both limited resources, plus natural gas and even conventional crude oil. These alternatives have been the best solutions to date at achieving all three of the *cheap*, *clean*, and *convenient* criteria.

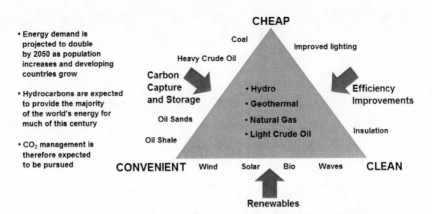

Figure 13.1. The trilemma. Note the trade-offs between cheap, clean, and convenient. (Adapted from: Royal Dutch Shell Oil Company, 2007)

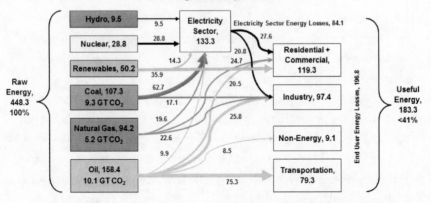

Figure 13.2. World energy flows and efficiency potential. Units of measure are EJ. Note that of the utilized raw energy, only 41% becomes useful energy to the consumer. Approximate calculations based on data from IEA, plus energy balances of non-OECD countries for 2002–2003. (Adapted from: Royal Dutch Shell Oil Company, 2007)

CO_2 Emissions by Fuel Type

Pounds per Billion BTU of Energy Input

Pollutant	Nat.Gas	Oil	Coal
Carbon Dioxide	117,000	164,000	208,000
Carbon Monoxide	40	33	208
Nitrogen Oxides	92	448	457
Sulfur Dioxide	1	1,122	2,591
Particulates	7	84	2,744
Mercury	0	0.007	0.016

Table 13.3. CO_2 and other emissions by fuel type. (Adapted from: EIA—Natural Gas Issues and Trends, 1998)

STEVEN E. SONDERGARD

There is a reasonable belief that a potential 50% general improvement in energy efficiency is possible with currently available technologies, bringing the overall useful energy from the 41% indicated in Figure 13.2 to over 60% of the raw energy. However, that improvement would only get part of the way toward the desired target of one-fourth of the CO_2 emissions per unit of energy. The noted decline in some cleaner burning resources (such as conventional crude oil) will make this target even more difficult to achieve.

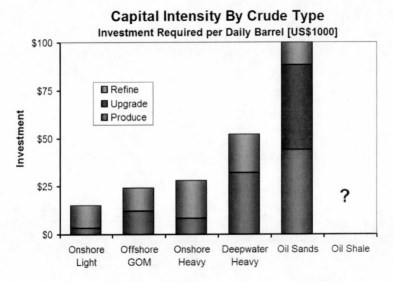

Figure 13.4. Capital investment intensity for various crude oil types. (Adapted from: Royal Dutch Shell Oil Company, 2007)

Potential Action Measures and Cost Considerations

One of the best approaches to date for understanding potential action measures is the climate change report by Vattenfall (a Swedish power utility). Their global cost curve of GHG abatement opportunities relative to business-as-usual has many valuable insights. Their curve was reproduced within a special report in *The Economist* on June 2, 2007.[2] Business-as-usual is defined at a WTI crude oil price at $40/barrel and natural gas at \$7/mcf.[3] However, the average future crude oil price will almost assuredly be higher than this, so the zero line on the graph would need to be shifted upward, maybe even significantly. Even without this possible change, the approach is still quite enlightening.

As shown in Figure 13.5, the life-cycle cost of each potential measure in euros per tonne of CO_2-eq is on the vertical axis, with the zero line at business-as-usual, and the horizontal axis is the accumulated impact of abatement in giga tonnes (billion metric tons) of CO_2-eq per year. The bars are the many action measures that could reduce CO_2-eq emissions, and they are arranged in order of economic life-cycle costs; negative cost simply means there would be a positive economic gain for implementing the measure. It is worth noting that this graph does not include changes in human responses, because history indicates that we humans do not like to change our habits for the most part, particularly as they pertain to creature comforts.

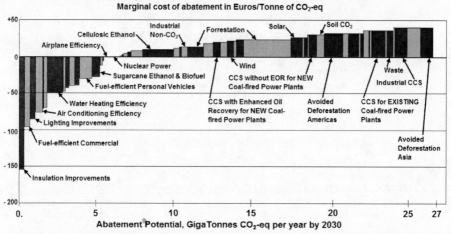

Figure 13.5. Potential abatement measures and costs. (Adapted from: Vattenfall; Cleaning up, a special report on business and climate change, 2007, *The Economist*, June 2, p.5)

STEVEN E. SONDERGARD

One of the first observations from Figure 13.5 is that there are multiple GHG improvement measures that are already economic and would increase GDP (lower left, below the zero line). If these actions were to be implemented, they would benefit society in more than one way: they would not only help reduce GHG emissions, but they would improve our economies as well. According to the figure, a total of all of the abatement measures that are currently economic sum to almost 7 giga tonnes of CO_2-eq.

A second observation from Figure 13.5 is that most of these measures with positive economics are energy efficiency projects. These include improved or additional insulation in homes and office buildings, improvements in commercial vehicle fuel efficiency, changing incandescent lightbulbs to fluorescent lightbulbs, using newer LED lighting, improving the efficiency of air-conditioning, efficiency improvements in water heaters, and improving passenger vehicle fuel efficiency. Using ethanol from sugarcane is another measure with sound economics, but some countries actually inhibit this action with an import tax on imported ethanol. The U.S. has a 54¢/gal import tax on ethanol.

If these action measures are already economic, why aren't these being implemented already? The answer appears to be multi-faceted, but we can reasonably speculate as to why this is the case. Some inaction might be due to the initial economic costs that prohibit action by either individuals or corporations. An example of this is the fact that the capital required to purchase extra insulation may not be available to some homeowners, or the hassle of installing insulation may limit taking action. Some inaction may simply be due to indifference. For example, a consumer may not believe Global Warming is *real* or that they could or should do something about it, even if it *is* real. Some inaction may be simply due to lack of knowledge. An example of this can be found when shopping for a new flat screen HDTV. One TV could easily use twice the power as a comparable product of similar screen size, but power consumption is not made readily available to the consumer.

Individuals will need to be a significant part of the solution. Yet, the average residential consumer cannot reasonably know what to do if they do not understand which appliances contribute what portion of the cost to their home utility bill. The largest improvements in energy usage by individuals would be obtained if homeowners tackled the biggest and easiest energy users first. Space heating and cooling is usually the largest energy consumer in a U.S. home, averaging around 55% to 60%. Installing thermally efficient air conditioners and furnaces will therefore be critically important. All appliances and home electronics are often the second largest user of household energy, averaging about 20% to 25%. Water heaters account for 14% up to 25% of the average home's energy. Note that new tank-less water heaters are estimated to have an energy savings of around 20% over the old tank types. When old hot water tanks fail, consumers should consider changing to tank-less water heaters.

Much of the total U.S. household energy is delivered in the form of electricity. Lighting accounts for about 15% of U.S. household electricity use. Installing fluorescent lightbulbs or light emitting diodes (LEDs) could reduce this dramatically. The DOE estimates kitchen appliances account for about 27% of average U.S. household electricity.[4] The refrigerator-freezer accounts for about 17%, TV/stereo/electronics account for about 15%, laundry accounts for about 13%, cooking appliances account for about 7%, dishwashers account for about 3%, and then there are some small miscellaneous users. Refrigerators are usually the largest electricity user in the home, but newer models use about a third of the electricity as thirty years ago. As far as cooking appliances go, induction stoves are the most efficient, followed by radiant ceramic and electric coil, then by microwave ovens, and then gas stoves, followed by electric ovens and finally gas ovens. It may be surprising that gas is so inefficient, but this is because gas utilizes only about 40% of the energy available to heat the pan and subsequently the food. It is much worse if there is a pilot. However, what isn't considered in these cooking appliance efficiencies is that average electricity generation from public utilities is only 33%–35% efficient, and most electricity is generated from high-CO_2-emission coal. Further, electricity has large transportation losses. In addition, if a natural gas stove is used at a time when space heating is required anyway, then its thermal efficiency can approach 100%. This is because the combustion products mix with the inside air and thereby heat the interior space as well, reducing the space heating requirements. The lesson is to always look at the bigger picture and be careful in how facts are presented or interpreted.

Figure 13.5 also shows measures that are marginally economic. These are those measures that are just above the zero line. These measures include airplane efficiency improvements, nuclear power generation, cellulosic ethanol, livestock and soil management, forestation, CCS with enhanced oil recovery, CCS for new coal power plants, industrial feedstock substitution, wind energy, and co-firing biomass. According to the figure, a total of all of the abatement measures that currently have marginal economics sum to *another* 7 giga tonnes of CO_2-eq. This means that 7 + 7 = 14 giga tonnes of CO_2-eq emissions per year are either already economic or have marginal economics. Most of these measures might even be economic at the expected higher oil prices, because the figure was constructed at \$40/bbl crude oil.

In a third category are measures that simply will not be economic without some significant help or breakthrough, as depicted in the upper right on Figure 13.5. These measures include re-forestation, solar energy, new coal plants using CCS, soil management measures, avoiding deforestation in America, retrofitting old coal plants with CCS, IGCC with CCS, industrial CCS, avoiding deforestation in Asia, and changes in waste management. A total of all of these abatement measures, which are unquestionably uneconomic, sum to approximately 13 giga tonnes of CO_2-eq, according to the figure.

A further observation from Figure 13.5 is that there are many measures that must be successfully pursued in order to have a sizeable dent in global emissions. None of the action measures are large enough by themselves to have more than minimal impact. This includes bio-fuels such as cellulosic ethanol, improved vehicle efficiency, nuclear, solar, wind, forestation, and even CCS. The reason is primarily because the scale of fossil fuel combustion is so large. A couple of examples might help clarify the scale. First, if 10% of energy sources were replaced by nuclear power, it would require 1,000 new nuclear power plants, and yet it might only slow warming by 0.2°F per century.[5] Second, if every car in the world were to average forty-three miles per gallon, it might only slow warming by 0.05°F by 2100.[6]

A related observation is that the accumulated impact of all measures up to about 40 euros/tonne CO_2-eq (over U.S. \$50/ton CO_2-eq) total only about 27 giga tonnes of CO_2-eq abatement. The world currently (2007) emits about 44 giga tons of CO_2-eq per year. The world is on a trajectory to emit about 58 GT/yr (per IEA) to 62 GT/yr (per Stern) by the year 2030, or by 2050 as much as 84 GT/yr (per Stern) to 80–90 GT/yr—ratioing GHG emissions to forecasted global GDP growth and assuming business-as-usual. A total improvement of 27 giga tonnes of CO_2-eq abatement is thus projected to be far short of stabilizing GHG emissions, and it is far from meeting any popular reduction target necessary to stabilize global temperatures.

The best way to attack the measures across Figure 13.5 appears to be by selectively targeted strategies. Measures that are already economic are unlikely to be made much more attractive with further economic incentives. Some benefits might be realized if consumer appliances were required to have labeling to inform the buyer how much energy is used by a specific appliance and how the specific appliance compares to other similar appliances. But many of these economic measures might just need to be regulated or mandated to be widely achieved. For example, governments could ban the sale of incandescent lightbulbs and require either fluorescent bulbs or—even better—new LED lighting. Measures that have marginal economics are mostly business driven, and they would be prime candidates for either government incentives or cap-and-trade programs to drive businesses toward taking action. The remaining measures that are currently uneconomic may need significant government support. Step changes in technology can be helped via government backing. Plus, coordinated international policy will be needed for such measures as country-specific forestation measures.

Slicing the data from Figure 13.5 yields some additional insights. Of the ~27 total giga tonnes of CO_2-eq abatement measures, the breakdown by sector is as follows: power (5.9), industrial (6.0), transportation (2.9), buildings (3.7), forestry (6.7), agriculture and waste (1.5). The breakdown by country is as follows: U.S. + Canada (4.4), OECD Europe (2.5), eastern Europe including Russia (1.6), other industrialized countries (2.5), China (4.6), other developing countries (11.1). From these breakdowns another observation is uncovered that the industry and power sectors represent less than 45% of the total reduction potential. Individual consumers must therefore participate extensively and must feel significant pain. Finally, it is observed that developing countries, including China, represent ~60% of the total reduction potential (>40% excluding China). These countries must be brought into the overall effort, or any endeavor cannot succeed. It can be concluded that the main focus of industrialized countries should be on energy efficiency improvements, plus the long-term economic support for technology innovation.

One of the few disappointments of the Vattenfall work is the apparent assignment of unrealistic probabilities to many action measures. The study assumes countries will work together in perfect coordination. It also assumes that all countries will favor policy and behavior that will reduce GHG emissions over economic development or any other competing factor. But a reasonable assessment indicates that full participation, particularly by individuals, to meet prescribed targets with full compliance of regulations appears quite unrealistic. In addition, the Vattenfall study assumes industry will not naturally relocate to countries where regulations are less stringent. This appears unrealistic as well.

To summarize this section, the action measures to counter Global Warming will necessarily be numerous. The envisioned potential action measures will not be sufficient to achieving the full results projected to be needed. About one-fourth of the envisioned potential GHG reductions are energy efficiency measures that are already economic. Much of the improvement from these measures will need to be by individuals rather than corporations, and they will likely need to be mandated. About one-fourth of the envisioned potential GHG reductions are industry measures that are marginally economic, and they could become economic with a carbon tax like cap and trade. The remaining one-half of envisioned potential GHG reductions will need significant R&D and government support. Further, the effort by developing countries including China and India needs to be roughly 60% of the potential GHG reductions, and the efforts of individuals will need to be over half of the potential GHG reductions.

Aspects of Policy Options Including Cap-and-Trade Measures

There are basically three ways for governments to develop policies to limit carbon emissions: 1) regulations or standards, 2) subsidies, and 3) putting a price on carbon. First, regulations could be used for abatement measures that already have positive economics. Regulations could be established for improving or adding insulation to homes and office buildings, for improvements in commercial vehicle fuel efficiency, for replacing incandescent lightbulbs with fluorescent lightbulbs, for efficiency improvements in heating and air-conditioning, for requiring the use of LED lighting, for efficiency improvements in water heaters, and for improving and encouraging passenger vehicle fuel efficiency.

The option of subsidies could be inefficient if it requires government to pick one technology over another technology. History indicates that governments have not been too good at this, and it is generally better to allow the best future technologies to be determined based on economics. Some people might point to corn-based ethanol as an example of this inefficiency.

The third option of putting a price on carbon is often considered to be the best of the three government options. It can be accomplished in two ways: by a tax or by a cap-and-trade system. Either of these mechanisms should *not* be considered as the sole means of control. A tax would set a price for emissions and would let the economic system determine the volume that would be economically emitted. A tax has the benefit of providing more certainty on the price or value of CO_2 and thereby provides reduced risk to industry. Further, a tax is considered far less costly to implement and is viewed as more bureaucratically efficient than cap and trade.

A cap-and-trade system would set the amount of CO_2 emissions desired and let the economic system determine the price required to meet it. The proposed cap-and-trade mechanisms are usually modeled after the successful acid rain mechanism from the U.S. Yet anyone with a modest understanding of chemistry should realize that a successful implementation of a cap-and-trade policy for carbon is *not* the same as the acid rain problem. Acid-rain-forming SO_2 is a peripheral product of combustion contaminants, while CO_2 is *the* product of combustion. Reducing CO_2 with any scale would effectively require the dismantling of the current economic engine of modern society. In essence, it would be a new world order. It is obvious to say that doing this prior to having an "adequate" replacement would be at least detrimental and quite possibly catastrophic for civilization. The term *adequate* has technical, economic, and scale components.

Either a tax or a cap-and-trade policy applied broadly across all measures could distort the market and does not allow a targeted approach for pursuing action measures. Either measure applied broadly assumes economics alone will drive action in every case. This assumption may be reasonably valid for business applications, but it appears that it would be less than effective for the action measures by individuals. A cap-and-trade system pursued alone may be more prone to giving individuals the false perception that they do not need to be a big part of the solution. Arguments are also made that a cap-and-trade system is a mechanism that primarily transfers wealth from industrialized countries or regions to developing countries or regions. In effect, a CO_2 tax and a cap-and-trade system are both a tax. It may not be obvious, but the costs from either mechanism would likely get passed on to the consumer.

A question of considerable debate for either a carbon tax or a cap-and-trade system is still whether governments can set a carbon price high enough or a target level low enough that it will make a meaningful difference in climate change and yet will not derail our society or our economies. Some form of cap-and-trade policy appears politically inevitable, but many contend it could easily derail the economy if great care is not taken in its application. Thinking ahead, what would happen to capping and trading carbon if it becomes widely realized that controlling carbon is either not in our best interest or it is detrimental to society as a whole? A tax could more easily be altered or stopped. A low impact cap-and-trade policy could temper the bet and still be used to steer action. But an aggressive cap-and-trade policy could become an economic house of cards. Considering the facts and uncertainties of the climate change problem, heavy-handed policy and its corresponding countermeasures could arguably lean toward irresponsibility.

Some promote the aspect that renewable energy efforts will create new jobs. But the downside is rarely mentioned. Job destruction could also occur, and net GDP could suffer when true economics do not determine job creation. A carbon tax or a cap-and-trade system could have more severe economic consequences than most people realize or admit. Care must therefore be taken that unmentioned downside impacts do not spoil good intentions.

Rich countries are often requested to take the first step toward

STEVEN E. SONDERGARD

reducing emissions. This doesn't seem unreasonable. But it should be realized that this is asking quite a lot of these countries, as it puts the economies of these industrialized countries at a considerable disadvantage until a point in time that all countries are treated equally—a condition for which a probability can be assigned, and some conclude it is an outcome unlikely to ever happen. By its very nature, the problem is a complex global issue. If the entire global community does not embrace the problem and countries are not treated equally in reaching the solution, not only does the problem remain unresolved, but the foundation of the economies upon which modern society is based erodes; and the world would arguably be in a much worse state than if no action were taken at all.

Another unfortunate reality with government policy is that current regulations often inhibit voluntary actions by companies that desire to go beyond current emission regulations. There would be many of the best companies who would gladly take the lead on the climate change front. Yet, the competitive driving forces of business are either not understood by those who set policy or are seemingly ignored. If a company desires to do as much as it can to reduce emissions, it is soon brought back to the reality of future competitive disadvantage that would be realized by doing so. It would be truly great if policy makers would realize this fact. But, alas, this may simply be too much to hope for.

Further Discussion of Energy Efficiency Improvements

There are indeed a few action measures for which there is virtually no debate. One of these areas is the area where abatement measures are already economic. These measures are essentially all in the category of energy efficiency improvements. These measures make sense both environmentally and economically. These measures make sense for energy security and for the conservation of scarce energy resources. We need to be doing these things, and we need to get started sooner rather than later. It is also apparent that governments will likely need to mandate most or all of these action measures in order for significant action to be taken.

Improving Vehicle Fuel Efficiency

There is a strong correlation between light duty vehicle penetration and GDP. Flexible light duty vehicles allow much of modern life and business to ensue. However, this does not mean vehicle fuel efficiency could not be dramatically improved without negatively impacting GDP. It is reported that the ratio of oil consumption to global GDP has already fallen by 22% since 1995 and OECD countries are leading the way.[7]

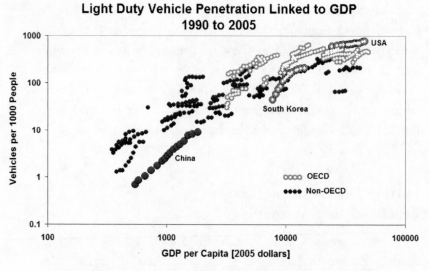

Figure 13.6. Vehicle penetration linked to GDP. (Adapted from: ExxonMobil, Energy Outlook, November 2007. By 2007 year-end, it was estimated that the U.S. had 800 vehicles per 1,000 people, Japan had 400 vehicles per 1,000 people, and China had increased to 25 vehicles per 1,000 people.)

STEVEN E. SONDERGARD

North American drivers in particular are relatively stubborn and spoiled. They have demanded performance, luxury, and size in their cars. This is partly due to the comparatively low gasoline taxes that have existed in the U.S. relative to other fuel-importing countries. This habit of vehicle size and poor energy efficiency will need to be broken, and it will assuredly need to be driven by a force large enough to sway people's actions.

About 10% to 20% of the emissions from hydrocarbon fuels come from the production, manufacture, and transportation of the fuels, while about 80% to 90% of the emissions come from the *use* of the fuel. In addition, the average North American automobile weighs nearly 4,000 lbs, while the average North American weighs less than 200 lbs, and most vehicles carry only one person most of the time. Therefore, most of the emissions from transportation fuels are a result of the work required to propel the vehicle rather than for propelling the occupant(s). Reducing the average vehicle weight plus reducing the wind resistance of automobiles could substantially improve the emissions per occupant-mile driven. The technology to achieve this already exists. Performance does not necessarily need to be sacrificed. Most cars sold in Europe and cars from most other countries are already smaller and lighter and have better mileage relative to the average North American car.

One method for attempting this mandate is an energy efficiency improvement measure to increase the U.S. CAFE standard for passenger vehicles. In support of this idea, the National Petroleum Council (NPC), in July 2007, unveiled a report declaring CAFE standards could double in the U.S. by 2030. Raising this standard is controversial, as it may hurt U.S. automobile manufacturers. While this may be true in the short term, improved efficiency is going to be required sooner or later. U.S. vehicles are already significantly behind foreign vehicles in regard to fuel economy. Yet, the automakers of U.S. vehicles know how to produce smaller vehicles because they produce smaller vehicles for foreign markets. The Mercedes Benz Smart Car has been sold in Europe for over a decade, but it is only now becoming available in the U.S.

It appears the automobile marketplace needs to have an external driving force applied to it in order to sway either automakers or North American customers to desire improved fuel efficiency and lower emissions. Why can't simple action steps be taken in the right direction? The lack of a large driving force appears to be the answer. If the automakers or consumers do not want to change course, a new tax might help sway the outcome. What about a tax on any purchased vehicle (new or used) that could vary proportionally to the vehicle's fuel economy? If a consumer desires to purchase a particular vehicle having an inefficient fuel rating, they could still do so and pay more tax. This might be similar to the luxury tax imposed some years ago. The proceeds from this tax might become doubly effective if they were to be given to the U.S. auto manufacturers that exceed their CAFE standard targets, with the automakers exceeding their CAFE standards the most receiving most of the money.

Another possible driving force for improving fuel efficiency might simply be a higher cost of fuel. But price-demand sensitivity for fuel is quite non-linear. Gasoline demand normally changes very little within a relatively wide price range. However, there is a price point at which the press, the market, and the pocketbook align to change public perceptions and initiate a change in habits. It is at this price point that a ratcheted step change can be observed to occur in which fuel efficiency considerations are elevated in priority. Although it is unfortunate that this type of step change in response is required before the majority of drivers change their habits, it has been repeatedly observed in recent history.

Potential Emission Reduction Measures from the Transportation Sector

Several improvement measures within the transportation sector have potential. The improvement measures include improved engine technologies, hybrid vehicles, lower emission fuels, aviation engine efficiency improvements, improved driving habits, fuel switching from gasoline to natural gas, fuel switching from gasoline to diesel fuels (diesel engines can get 30% more MPG than gasoline engines because diesel has a higher energy content than gasoline, and diesel engines currently have a higher engine efficiency versus current gasoline engines), second-generation bio-fuels (first-generation bio-fuels do not have lower GHG emissions), public transportation, and urban planning.

It can be reasonably expected that future technological improvements will occur for each of hybrid, gasoline, and diesel vehicles, and their life-cycle emission footprints will be reduced in the future. But it should be noted that both fuel density and fuel transportability considerations lead to the conclusion that liquid fuels will probably remain the most efficient transportation fuels. Liquid hydrocarbon fuels have already gone through significant changes to lower the emission characteristics of both gasoline and diesel fuels. For example, 97% to 99% of the sulfur in gasoline and diesel is already being removed from these fuels by oil refineries. While most of the potential fuel improvement has already been achieved, this trend to improving fuel quality can be expected to continue on an asymptotic trajectory.

Europe has already taken the step of transitioning its vehicle fleets to higher-efficiency diesels. This was originally a strategy initiated by France and Germany for their car companies to better compete with Japanese gasoline imports. European taxes and subsidies have driven vehicle sales towards diesel relatively quickly. However, there are two reasons why it will not be so easy for a change like this to occur in North America. First, refineries in the U.S. primarily convert their heavier crude fraction called gasoil within fluid catalytic cracker units (FCCUs), while European refineries perform this task primarily with hydrocracker units (HCUs). This difference results in a much better cetane diesel fuel in Europe than in the U.S. Completely rebuilding most U.S. refineries would be required for the same diesel quality. Second, Europeans have switched to diesel in such numbers that much of the gasoline naturally produced by

European refineries must be sent to the U.S. and Middle East just to find a home for it. If the U.S. were to adopt a similar dieselization strategy as Europe has pursued, there would be a world glut of gasoline relative to other refinery products. This would necessitate a further rise in the price of the other key products, such as jet fuel and diesel fuel, relative to gasoline.

Several recent developments in gasoline and diesel engine and fuel technologies have been examined by the Southwest Research Institute (SRI). These technologies follow, with subsequent efficiency and emission findings from SRI:[8]

- High efficiency dilute gasoline engines (HEDGE)
- Homogeneous charge compression ignition (HCCI)
- Controlled auto ignition (CAI)
- Premixed charge compression ignition (PCCI)
- Low temperature combustion (LTC)
- Premixed compression ignition (PCI)
- Compression and spark ignition (CSI)

Overall, engine combustion technologies are converging to some of the same general characteristics. These include:

- Delayed ignition and rapid burn rates
- High exhaust gas recycle
- Part or all of the fuel is premixed
- Lower NOx emissions
- Higher HC and CO emissions, prior to exhaust treatment

Peak engine efficiencies occur when combustion is phased such that the peak heat release rate occurs at $12°$ after top dead center with heat release durations of $10°$ to $30°$. Fuel aromatics delay combustion initiation (ignition), and n-paraffins tend to speed it up. Diesel engine data also suggests diesel cetane has little effect on NOx emissions, but cetane variability is an issue, such that controlling the cetane *range* will likely be required in the future. Fischer-Tropsch fuels and green diesel fuels have lower NOx, lower sulfur, lower PM, and higher HC emissions. Bio-diesel fuels have lower HC, lower CO, and increased NOx emissions. For HCCI technology, octane or cetane doesn't describe fuel quality, and fuel efficiency is improved by about 10%. For HEDGE technology, NOx emissions are significantly lower with CO and HC emissions equivalent to partial zero emission vehicles (PZEV) directly out of the engine, and thermal efficiencies have approached 39%. In mid-2007, HEDGE technology showed the most promise for gasoline fuels, and it only required an eighty-seven octane fuel. In summary, future gasoline engine technologies have the promise of significantly increased efficiency, and these engines are expected to rival future diesel engine technologies in efficiency. Additionally, high-efficiency gasoline technology has about $5,000 in savings on the vehicle purchase cost over high-efficiency diesel technology.

Further improvements in transportation fuels are expected as well. It is hoped that second generation bio-fuels will be able to have lower overall GHG emissions. This is indeed the expectation for cellulosic ethanol. Aviation (jet turbine) engine efficiencies have been on an improvement trajectory for some time, and this trend is projected to continue into the future. Increased public transportation has also been springing up across North American cities. The driving force for this trend has assuredly been propelled in large part by recent increases in fuel costs. Changes are also being observed in urban development where designs are starting to utilize smaller-community approaches with reduced requirements for travel. With the large distances between cities in North America, relative to Europe, community development is more challenging, but this should still be a goal of local and state governments. Improvements in urban development will require a continual and persistent effort to reduce miles driven.

Continuous improvement on each of these aspects of the transportation sector can be expected. But it is clear the technology already exists to provide huge potential improvements in the energy efficiency of personal vehicles and thereby huge reductions in GHG emissions. By 2050, a reasonable estimate might be that the entire transportation sector could see a reduction in GHG emissions of more than 50%, relative to current emissions. That equates to an improvement in energy efficiency of over 100%.

Hybrids and Plug-in Hybrids

Electricall powered vehicles can have lower life-cycle emissions than conventional gasoline vehicles, even though they have about twice the emissions in the manufacturing process. The emissions from manufacturing hybrid vehicles lie between those of electric and gasoline technologies, but emissions from operating these vehicles can be about half that of current gasoline technologies. Toyota, in particular, has made great strides in developing hybrid vehicles. Current gasoline engine hybrids are lowering life-cycle emissions from 5% to 40+% relative to gasoline-only vehicles. Note that several of the new gasoline and diesel technologies described in the previous section are expected to have roughly equivalent GHG emissions compared to today's best hybrids.

Plug-in electric hybrids have even more potential at reducing GHG emissions than gasoline hybrids, and the batteries to power these vehicles are receiving high research and development focus. China and Korea are currently several years ahead of the rest of the world in this effort. The most promising battery technologies use lithium, as it is lighter in weight. Bolivian salt flats also contain enough lithium to supply the world with batteries for decades.

Plug-in electrics are zero emission vehicles. But this only applies to their operation. Consideration must also be given to the efficiency and GHG emissions generated by the electric utility plants. The average electricity generation by public utilities in the U.S. is only 33% efficient, and 50% of electricity generated in America is from high-CO_2-emission coal. Recall that new-generation gasoline engine technology (HEDGE) approaches 39% thermal efficiency. Therefore, as amazing as it seems, gasoline powered vehicles using currently available technology would be more efficient overall and produce less overall GHG emissions than zero-emission plug-in electric vehicles. This would hold true until the average efficiency of the electrical power generation improves significantly and/or the average GHG emissions from these plants are reduced substantially. This will take many years of transition, and possibly decades. Further, the automobile batteries alone for plug-in electric vehicles can cost more than $20,000, and they have a range of only 60 to 100 miles between recharges. Gasoline would need to be about $8/gallon before these electric cars would be economic. Electric hybrids and electic vehicles also have greater post-life and disposal concerns.

Renewable Energy

Renewable energy includes biomass, wind, solar, hydroelectric, and geothermal. All renewable sources currently produce about 18% of global electricity supply and 13% of the world's energy needs.[9, 10] Biomass alone provides about 10% of world energy, and most of that is in undeveloped countries where animal manure is burned as the fuel in simple home stoves. Bio-fuels currently represent only about 1% of the global transportation fuel market. Ethanol from both corn and sugarcane is rapidly growing. Hydroelectric power may be approaching its limit in the industrialized world, but in developing countries it is still expanding, particularly in China. Per the USDOE/EIA, renewable energy was 5.5% of U.S. energy consumption in 2007, with 2.2% from ethanol and bio-diesel. China currently meets just under 8% of its energy needs from renewable sources, most of it from hydroelectric. China announced in 2005 that by 2020 it expects to generate some 15% of its energy needs with renewable sources.

Globally, the growth of renewable fuels from 2005 to 2030 is expected to be about 1.5% per year according to ExxonMobil (EIA projected 2.5% per year in 2007). Wind is expected to grow at 10.5% per year, solar is expected to grow 9.9% per year, bio-fuels are expected to grow at 7.6% per year, hydro/geothermal is expected to grow at 2.0% per year, and biomass is expected to grow at 0.7% per year.[11] This relates to a projected overall global energy demand growth at 1.3% per year. Even with stagnant global energy demand and a strong push to double or triple renewable energy, the scale of the overall global energy usage prevents renewable energy from sizably penetrating the global energy mix for many years to come.

It is apparent after some study that most fuels generated from biomass currently do not have lower emission levels than normal hydrocarbon fuels. These first-generation bio-fuels have generally been a disappointment. In fact, some of these bio-fuels can have several times the life-cycle emission levels as normal hydrocarbon fuels, depending on the feedstock and the deforestation required to make the fuel. The hope is now in second-generation bio-fuels. Unfortunately for bio-fuels, the scale of fossil fuel demand is so great that this alone limits the impact of bio-fuels. Bio-fuels have been promoted in the U.S. now for over twenty-five years, and they still have only a combined market penetration of between 2.5% and 3.5% in the U.S. today. The dominant bio-fuels are described below.

Ethanol

Ethanol is touted as a fuel of the future mostly because it is renewable. The Pembina Institute and Argonne National Laboratory indicate ethanol can have reduced GHG emissions if smart decisions are made in its production. The University of California (Berkley) concluded recently that a gallon of ethanol has about 10% to 15% lower GHG emissions than a gallon of normal hydrocarbon motor gasoline. There are about 19.4 pounds of CO_2 produced in the combustion of a gallon of normal hydrocarbon gasoline. This author concurs with these assessments as well. However, it must also be understood that ethanol has only about 70% of the energy content per gallon compared to normal hydrocarbon motor gasoline. Therefore, on the basis of the combustion energy available, ethanol has about 10% to 20% *higher* GHG emissions than normal hydrocarbon motor gasoline. To put it in perspective, this approaches the calculated GHG emissions for oil sands. It is ironic that fuel sources with similar GHG emissions are being perceived so differently. Yet, if any land is cleared in order to plant the corn feedstock, emissions from ethanol can get much worse. To corroborate, an August 2007 report in *Science* stated bio-fuel "is a mistake in climate change terms" and "has between two to nine times more carbon emissions"—the higher number resulting when forested areas are cleared to grow a crop for producing the fuel. Some recent reports indicate corn-based ethanol has only 60% of the life-cycle CO_2 emissions compared to normal hydrocarbon gasoline. These reports that indicate such low life-cycle CO_2 emissions try to account for the fact the corn crop itself captures CO_2 from the atmosphere and thereby offsets the emissions from the combustion of the ethanol. However, this is rather misleading. These analyses are flawed in that their mass balance envelopes are drawn inappropriately. If corn were not grown for ethanol, then either corn would be grown for food or an alternate crop would be grown on the land—and these alternate crops would capture CO_2 as well. In addition, these alternate crops would be utilized for something other than being combusted to CO_2, such as food or clothing. So, the life-cycle CO_2 emissions for corn-to-ethanol should *not* consider the CO_2 absorption of the corn crop alone.

Unfortunately, ethanol from corn is not very energy efficient either. This means that corn-based ethanol does not appreciably improve energy security or reduce crude oil imports. It takes about 75–85% of the energy to make corn-based ethanol than is available in the ethanol itself (assuming a modern ethanol plant using molecular sieves to remove the water from the ethanol and not older less-efficient solvent-extraction technology). This energy efficiency does vary between plants, and it depends on how much heat integration an ethanol plant incorporates in its design and its operation. Plant efficiencies have generally been getting better over time. It also helps that modern ethanol plants have yields that have improved to typically produce 2.8 gallons of ethanol per bushel of #2 dent corn. There are also those who claim ethanol production uses more energy than is available in the ethanol itself. But these claims usually do not consider the energy value of the animal feed byproduct dried distillers grains (DDG) in their calculations.

For comparison, ethanol from Brazilian sugarcane requires only about 14% to 20% of the energy input available in its ethanol product, and its life-cycle CO_2 emissions are about 20% of normal hydrocarbon gasoline. The main reasons for this four- to sixfold difference include: sugarcane has a higher sugar content than corn, Brazilian sugarcane can get up to three harvests per year, sugarcane does not need planting every year like corn (the last two points decrease the plowing and planting energy relative to corn), Brazilian sugarcane often does not require extra irrigation, sugarcane requires less fertilizer than corn, and the remaining biomass material (known as bagasse) from the crushed sugarcane is utilized in the local Brazilian communities as fuel. Ethanol from sugarcane is so significant in Brazil that it currently accounts for about 40% of the fuel used in the cars within that country. Additionally, leveling the economic playing field by removing import taxes on ethanol from sugarcane would benefit the environment.

Ethanol from corn is also a trade-off with our world's food supply, as approximately 24% of the U.S. corn crop went toward ethanol production in 2007. About 30% of the 2008 corn crop was projected to go toward ethanol production. This is actually scary to some, because the world is already on the edge of its ability to produce the food required to feed its population. An increase in the price of #2 dent corn causes the price of sweet corn, wheat, barley, and other grains to also increase because farmers plant less food crops in order to maximize their profit. With only limited crop land available, the planting of more #2 dent corn must necessarily reduce the planting of other food crops. Further, #2 dent corn is used to feed livestock, which results in higher prices for meat, milk, and eggs. This domino effect elevates the price of many agricultural products.

Several opinions of food price sensitivity follow. According to a study by Iowa State University, food prices increased by $47 per person in the span of a year as a result of corn-based ethanol alone. A U.S. Labor Department report of August 2007 indicated food prices went up an average of 4.1% from July 2006 through July 2007, even with some prices going up much more than that. The International Monetary Fund (IMF) has indicated average world food prices went up 10% in 2006 alone. The Agence France-Presse reported that they expect food prices to further increase by 40% to 80% due to forecasted increases in renewable energy (predominantly from corn). The Congressional Budget Office (CBO) estimated that increased ethanol use contributed to the 50% increase in the price of corn and accounted for 10% to 15% of the 5.1% increase in all food prices from April 2007 to April 2008. The World Bank stated in 2008 that the price of food *staples* had jumped 80% since 2005, and 70% to 75% of this increased price in these food staples was due to biofuels. Energy and fertilizer are reported to account for just 15% of that increase. Alternatively, the Renewable Fuels Association argues the price of fuel should have impacted the food prices more than the ethanol production from corn. In early 2008, corn was hovering around $6/bushel (this was an increase of about $2/bushel for each of 2006 and 2007). It should be noted that the U.S. grows about 40% of the world's corn, but it accounts for about 70% of world exports. A reduction in U.S. corn exports will therefore impact poor countries and poor people much more than the average.

In July 2007 there were 119 operating ethanol plants in the U.S., yielding a total of 6.5 billion gallons per year of ethanol. Eighty-six ethanol plants were under construction at that time. In 2008 there was over 9 billion gallons per year of ethanol capacity in the U.S. Note that the Renewable Fuels Act mandated 7.5 billion gallons per year by 2012. In late 2007, the Renewable Fuels Standard (RFS) was enacted, calling for increasing ethanol production in the U.S. by roughly five times to 36 billion gallons per year by 2022, with 15 billion gallons per year supplied by corn. This would be more than double the requirement from corn in 2007, and it is about 50% of the entire 2007 U.S. corn crop. But what impact will this have on food prices? To make matters worse, the world demand for food is expected to double by 2050. Because of the trade-offs, predominantly between fuel and food, many believe a market-driven mandate for ethanol would be better than a hard government-driven ethanol mandate because it would allow for more flexibility in preventing a catastrophe if this mandate proves to go too far in producing fuel at the expense of food.

There will also be unintended, but predictable, consequences from increased ethanol production. One consequence is that there will be price increases for jet fuel and diesel fuel. This may not be obvious, but this conclusion is realized with cause-and-effect analysis. Ethanol will probably be looked at to help reduce the dependency on crude oil even when the gasoline supply is long. In this environment, increased ethanol production will necessarily further reduce the demand for hydrocarbon gasoline, thus dropping the relative price for the hydrocarbon gasoline. The gasoline price would actually be expected to drop for *any* mechanism that reduces gasoline demand. U.S. refiners would then be forced to increase the price of other products in an attempt to make a profit. The products that are in the shortest supply can be expected to bear the brunt of the relative price increases, and these are currently jet fuel and diesel fuel. The predictable result will be increased economic stress on the airline and trucking industries, and that would ripple further consequences across our economy and society.

It should be noted that ethanol is not the perfect gasoline substitute. It readily absorbs water, and it is a strong solvent. Further, ethanol has a pure vapor pressure of only 2 psia at 100°F (it is low because

STEVEN E. SONDERGARD

it is so polar). Ethanol is also chemically dissimilar to gasoline, and when it is blended into gasoline at 10% by volume, it blends linearly as if it has a vapor pressure of 17 to 19 psig—which is significantly higher than normal summer specification gasoline, currently around 7.8 psig, depending on the state or region. The result is a higher vapor pressure gasoline that translates into more volatile organic compound (VOC) emissions. In addition, ethanol has a high octane value with an (R+M)/2 of 110, but its blending octane (R+M)/2 is only about 97.5. The actual mileage obtained from ethanol use in automobiles is not solely a function of its energy content relative to hydrocarbon gasoline, but it also varies with the combustion characteristics of the engine, how the engine is tuned, the engine temperature, how clean the engine is, and the absorbed water content of the fuel, among other factors.

Ethanol is already partially oxygenated, and its combustion results in increased ozone and formaldehyde emissions relative to normal hydrocarbon gasoline. In contrast, its combustion is known to have reduced carbon monoxide emissions. Economically, ethanol is currently subsidized by the U.S. government at 51¢ per gallon, not counting some small plant subsidies and some additional state subsidies. Without government mandates and subsidization, corn-based ethanol plants would not be economic on their own. The recent run-up in crude oil prices has helped the economics of corn-based ethanol plants tremendously, but the counter run-up in corn prices and lower DDG prices have depressed the economics again below hurdle rates such that the construction of new ethanol plants currently have generally unfavorable economics at this writing.

In addition, ethanol plants using corn utilize a large quantity of water, and this is often a resource in short supply within many farming communities. Not only is the corn usually irrigated from which ethanol is made, but ethanol plants themselves use an additional three gallons of water for every gallon of ethanol produced. Mid-continent U.S. aquifers, such as the Ogallala aquifer, are declining rather rapidly. Realize that ethanol produced from corn is only renewable while ample water is still available.

In summary, ethanol produced from corn may not be the best policy. The ethanol fraction of gasoline alone does not determine the crude oil savings. Crude oil produces only about 50% gasoline. Ethanol is currently blended into gasoline at no more than 10%. Corn ethanol was described earlier as having about 20% energy savings versus normal hydrocarbon gasoline. Therefore, if all gasoline sold contained maximum ethanol, then there would still be only 50% * 10% * 20% = 1.0% savings in equivalent crude oil consumption. This is a key point. Some may suggest this is better than nothing, until the impact on our food and water supply is considered. We should be able to do better. Ethanol from sugarcane is already economic and environmentally beneficial. Ethanol from non-food feedstocks such as cellulosic or waste materials could be the best option for producing ethanol, but help will be needed from multiple technology and cost-reduction breakthroughs.

Cellulosic Ethanol

There is significant effort with some promise to soon be able to economically make so-called cellulosic ethanol from ligno-cellulosic materials, such as newspapers, wood chips, switch grass, etc. The predominant problem basically lies in breaking down and converting the C_5 sugars from the hemi-cellulose molecules. This breakdown does not occur naturally. Yet, there are several known ways this can be done, including strong acid, weak acid, gasification, and enzyme techniques. Each of these processes is technologically feasible today, but none of these processes are very efficient, either thermodynamically, biologically, or economically.

What is believed to be the first commercial scale cellulosic ethanol plant in the world had its ground breaking in November 2007. This new cellulosic ethanol plant is being built in Georgia by Range

STEVEN E. SONDERGARD

Fuels, using waste products from wood. It is also being massively subsidized by the USDOE. The initial construction cost for 20 million gallons per year was $225 million with a $76 million government subsidy. A year later, the cost escalated to $326 million with further government support. This plant is expected to convert wood chip feedstocks that cost about $20/ton delivered to a synthetic gas and then recombine the gases to ethanol. The energy utilized in this process is expected to be significantly lower than the energy used for corn-based ethanol, using only about 20% to 25% of the energy content of the ethanol product itself.

Another cellulosic ethanol company, ZeaChem, is scheduled to build a (smaller) demonstration plant in Oregon, with startup targeted for 2010. This facility is expected to produce about 1.5 million gallons per year at a cost of about $40 million. This process is expected to produce about 135 gallons of ethanol per dry ton of biomass, versus about 90 gallons per ton from other processes. This process uses a common bacteria found in termites in a bio-thermo-chemical reaction.

Over the last twenty years or so, economic cellulosic ethanol has been forecasted to always be five to ten years away. Today there is still much hope that it will become a reality, and a lot more effort and capital is being utilized to make it happen. The U.S. government has big plans to support this effort by mandates and subsidizations, and legislation has been devised that makes a huge bet on the eventual economic commercialization of cellulosic ethanol production. The current tax credit alone for cellulosic ethanol is 50¢/gallon greater than corn-based ethanol, or $1.01/gallon. But do not expect cellulosic ethanol to become widespread within less than a decade or maybe two. It will take time for breakthroughs to occur, for processes to be improved, for permits to be obtained, for designs to be completed, for construction to occur, and then for significant quantities to become produced. Potential feedstock quantities on the scale needed to supplant hydrocarbon gasoline will also limit the possibility that cellulosic ethanol will displace a majority of hydrocarbon gasoline. The hope that cellulosic ethanol may become a large part of the near-term Global Warming solution may therefore be unwarranted.

Bio-diesel, Green Diesel, and Renewable Diesel

Bio-diesel is defined by the National Bio-diesel Board as a fuel comprised of mono-alkyl ester of long chain fatty acids derived from vegetable oils or animal fats, meeting the requirements of ASTM D6751 and designated as B100. The international standard is EN 14214. These are standard laboratory test procedures that are designed to ensure product quality. These oils and fats are triacylglycerols, sometimes called triglycerides, and can come from many sources. Soybean oil is the most common feedstock in North America, and palm oil is the most common feedstock in the Far East. The fatty molecules are composed of a glycerol core with three long-chain branches called fatty acids. There are a series of simple steps to convert these fatty molecules to bio-diesel. The main step is called *transesterification*, where the fat or oil is reacted with an alcohol, usually methanol or ethanol, in the presence of a strong alkaline catalyst, usually sodium hydroxide or potassium hydroxide. This step replaces the glycerol with the methyl or ethyl alcohol and results in the products of glycerol (a byproduct at ~10 wt%) and mono-alkyl esters (the bio-diesel). The subsequent steps include the removal of the glycerol, removal of the catalyst, and removal of un-reacted alcohol.

Bio-diesel is almost always blended into normal hydrocarbon diesel in various quantities. B20 designates a 20% blend, B5 designates a 5% blend, B2 designates a 2% blend, and so on. One of the main reasons for this is that B100 does not have good cold flow properties, specifically cloud point and pour point—it just does not flow very well in cold temperatures. The positive aspects of bio-diesel include improved lubricity, lower toxicity, higher flash point, and lower emissions, including CO, SOx, and some particulate matter (PM). The negative aspects of bio-diesel include poor cold flow properties, increased NOx emissions, increased PM2.5, lower fuel economy, and its solvent characteristics. Bio-diesel is somewhat polar, and it will therefore tend to absorb water, which can cause biological growth in the fuel and/or corrosion from the fuel.

The CO emissions are about 50% better for bio-diesel than normal hydrocarbon diesel. The PM emissions are about 45% better. The NOx emissions are about 10% to 20% higher. The SOx emissions are low because the fuel sulfur levels are low. The flash point is about 300°F. The cetane number is higher than most normal diesels. Depending on the feedstock, the cold flow problems can start to have an impact at 15°F up to as high as 70°F (when using animal fat as a feedstock). The fuel economy for bio-diesel is about 10% lower than conventional #2 diesel. The tailpipe CO_2 emissions are about 5% higher for bio-diesel. There are about 21.7 pounds of CO_2 produced in the combustion of a gallon of normal hydrocarbon diesel. The life-cycle CO_2 emissions are sometimes touted as 60% less than for normal hydrocarbon diesel. But life-cycle emissions depend greatly on what would have been grown on the land if the bio-diesel feedstock were not grown. The 60% number assumes absolutely nothing would have been grown on the land in question. Obviously if rain forests are cleared in order to grow palm oil, as is currently being done in Indonesia and the Philippines, the life-cycle CO_2 emissions would be much higher and multiple times higher than normal hydrocarbon diesel.

In September 2005, Minnesota became the first U.S. state to mandate B2 (2% bio-diesel) in all diesel fuel sold in the state. The 0°F winter days that were experienced in the subsequent winter resulted in many diesel fuel problems, which virtually shut down the trucking industry in the state. In December 2005, Minnesota lifted the mandate for fifty-one days to study and resolve the issues. The cold flow, bacterial growth, and glycerin separation/contamination problems that were encountered were eventually overcome prior to the subsequent winter by using better quality control measures.

Finally, the cost of bio-diesel is high and increasing due to the limited feedstock availability and the rising costs of its agricultural feedstocks. The global production of bio-diesel in 2005 was estimated at ~1.0 billion gallons (about 67,500 barrels per day). Europe is the largest producer of bio-diesel in the world, and Germany is the largest producer within the European Union. The 2005 U.S. production of bio-diesel was approximately 75 million gallons per year, which was about 5,000 barrels per day. In July 2007, there were ninety-six plants scheduled to be online within eighteen months in the U.S. to yield a total of 3.28 billion gallons per year. However, by late 2007, bio-diesel plants suddenly became uneconomic in the U.S. due to limited and high-cost feedstock. As of this writing, many U.S. plants recently built are simply shut down or for sale. The potential impact of bio-diesel is restricted by the limited availability of cheap feedstock, which is about four million barrels per day in the U.S., and the limited market for the byproduct glycerol. It is currently more profitable to convert waste vegetable oil into other products, such as soap. New uses for the glycerol are currently being examined and developed. If the economics eventually get resolved, the total potential for converting all vegetable and animal fats in the U.S. is estimated by some to be around 3 billion gallons per year, or just under 200,000 barrels per day.

Overall, first-generation bio-diesel is generally not considered an improvement over normal hydrocarbon diesel in GHG emissions, particularly if forests are cleared for the oil feedstock. The real promise for bio-diesel is in the potential for new feedstocks such as jatropha or even algae. Whereas soybeans have a converted yield of about 48 gallons per acre, jatropha has a converted yield of over 200 gallons per acre, and algae has a converted yield of 800 to 1600 gallons per acre. Experimental trials during DOE's National Renewable Energy Laboratory (NREL) aquatic species program showed the most promise for algae.

Green diesel and renewable diesel are terms used for diesel

STEVEN E. SONDERGARD

fuels that use the same feedstocks as bio-diesel, but the processing is completely different. These diesel fuels are the result of these same feedstocks being hydrotreated in a conventional refinery diesel hydrotreater. A separate reactor can be used to first take out the contaminants, such as oxygen, calcium, and phosphorus. The hydrotreating step uses a high pressure and high temperature along with a high hydrogen partial pressure over a fixed catalyst bed of cobalt-molybdenum or nickel-molybdenum. An isomerization stage can be used to isomerize the paraffins to improve the cold flow properties of the fuel. A separate processing train may end up being required to receive the tax credits. The product could be used as a high-quality jet fuel (it could even be used to produce JP-8) or alternatively be used as a high-quality diesel fuel. The product properties are virtually identical to Fischer-Tropsch (coal-to-liquids) product. The cetane is seventy to eighty, the cloud point is +15°F (65° to 70°F without the isomerization), the sulfur is <1 ppm, and the stability is good. Both Universal Oil Products (UOP) (green diesel, recently termed *Ecofining*) and Neste Oils NExBTL (renewable diesel) have the licenses for competing technologies. The processes have a liquid volume conversion around 100%, including 0–10% naphtha product; plus, there is some LPG and fuel gas produced.

Jatropha

Jatropha is a genus of about 175 shrubs and trees. *Jatropha curcas*, sometimes called physic nut, is the most promising species of the genus for oil production. These poisonous trees are thought to be native to the Caribbean and Central America but have become naturalized in India, Africa, and Mexico. The trees can grow to twenty feet tall, they can grow in harsh environments, they can survive up to fifty years, and they are resistant to droughts and pests.

They are the only plant currently known besides sugarcane to produce bio-fuel economically, without subsidization. The tree produces an inedible nut about the size of a golf ball that can be harvested with a universal nut sheller. The yield is variable, and the nuts don't all ripen at the same time. Harvesting twice a year appears to be common. The nuts/seeds can be crushed to produce an oily material. A couple of companies marketing the tree are reporting 600 to 1,000 gallons per acre. But the more accepted yield is closer to 200 gallons per acre. Long-term impacts on soil quality are unknown at this time. It is currently being developed in India for fuel oil. Jatropha is receiving strong interest in Indonesia, the Philippines, China, and Brazil.

Bio-butanol

Bio-butanol is simply butanol produced from biomass. The historical process for producing bio-butanol is the acetone-butanol-ethanol (ABE) process with the bacterium *Clostridium acetobutylicum*, known as the Weizman organism. The process is very similar to the fermentation process used to produce ethanol. The feedstock is the same as for producing ethanol, and existing ethanol plants can be economically converted to produce butanol. In addition to butanol, the process produces hydrogen, acetic acid, lactic acid, propionic acid, acetone, ethanol, and isopropanol. The production of butanol from the historical fermentation process is less than 2%, and it was separated from the other products by an energy intensive distillation and extraction process. It was therefore fairly costly to produce by this means.

Research on improving the process yield and reducing the energy intensity is being spearheaded by a partnership between British Petroleum, Dupont, and British Sugar, announced in June 2006. A higher conversion process is expected by maybe late 2008, with another generation of improvement by 2010. A 5000 gallon per year pilot plant is scheduled to start up in 2010. Newer fermentation processes are expected to double the yield from 1.3 to 2.5 gallons per bushel. For comparison, ethanol yields are roughly 2.8 gallons per bushel, but ethanol has a lower density.

STEVEN E. SONDERGARD

A competing next-generation butanol process is being developed by a company called Gevo. Their process development should be in the pilot plant stage by 2009. They use an altered E. coli bacterium to more selectively produce either 1-butanol or iso-butanol. The process can reportedly produce butanol at 99% selectivity, with projected yields of 2.3 gallons per bushel from corn feedstock. Gevo claims they could convert an existing ethanol plant for about $25 million.

Butanols contain about 40% more energy per gallon that ethanol, but they still have about 10% less energy than normal hydrocarbon gasoline. The emissions benefits of butanols are believed to be at least as good as ethanol in gasoline, and butanols can be used as an additive in diesel fuel as well. Butanols are less polar than ethanol, which gives it a good chance at being able to be transported in existing hydrocarbon product distribution pipelines. 1-Butanol has a flashpoint >140°F (>60°C), so it can be utilized in jet fuel. Greater quantities of butanol can be blended into gasoline than ethanol. 1-Butanol has a reasonable octane with (R+M)/2 of 87, while iso-butanol has a better octane with (R+M)/2 of 97. Butanols also have lower vapor pressures than ethanol (~1.3 psia at 100°F). 1-butanol has a flash point at 99°F, while iso-butanol has a flash point at 88°F. 1-Butanol has a boiling point at 243°F and a melting point at -130°F. Iso-butanol has a boiling point at 211°F and a melting point at -174°F. The viscosity of butanols are roughly the same as for normal diesel fuel.

Fuel from Algae

Algae actually have one of the most promising potentials to produce future hydrocarbon fuels. Algae have the capability to consume high quantities of CO_2 from the atmosphere, and they have a very high oil production capability. As mentioned earlier, algae has a converted yield of over eight hundred gallons per acre, and experimental trials during NREL's aquatic species program showed even greater promise. Because algae are a single-cell life form, they have high growth rates compared to plants. There are maybe a hundred thousand naturally-occurring algae species, and some of them could become virtual algal oil factories for fuel production, while also consuming CO_2 from the atmosphere as a feedstock or potentially consuming an industrial flue gas stream. The possible use of algae for fuel could help resolve several vexing problems at once.

Proprietary algae bio-reactors that stress the algae into producing more oil should be able to produce higher quantities of a hydrocarbon product in the boiling range for gasoline, jet fuel, or diesel fuel. Much of the raw crude-like oil can simply be pressed out of the dried algae. The hydrocarbon products produced from the algae could be further isolated and tailored to consumer demand with one or more of the existing refinery processes and technologies. It isn't hard to make a little oil from algae. The difficulty will be in the economic efficiency and the required scale. The estimated cost of algae-based fuels at volume using current technology is about six times that of conventional hydrocarbon fuels. The commercial viability for algae-based fuels is therefore estimated to be ten to twenty years away.

Wind, Solar, and Waves

Denmark currently generates about 20% of its electricity from wind.[12] This focus started after the 1973 oil embargo, and the result is that Denmark is now one of the world leaders in wind turbine research and construction. Spain is rapidly growing its wind energy as well. In 2000 Spain had only 2,200 MW of wind energy, but in 2008 Spain had increased that to 15,500 MW. In the U.S., wind generates less than 1% of the electricity, but it still amounts to about 25,000 MW. About 8,000 MW of this capacity was added in 2008 alone. Wind turbines keep getting bigger and lighter, and therefore electricity generated from them keeps getting cheaper. Turbine blades are now up to forty meters long, and they can produce up to 3.3MW per turbine at a cost between 7¢ to 8¢/kwh. This actually isn't too bad. But wind turbines with sixty-two meter turbine blades producing 5MW should be cutting costs further very soon. General Electric is working toward a ninety meter turbine blade design that they believe could produce electricity for 4¢ to 5¢/kwh (but increasing steel, copper, and labor costs might increase those numbers). Other manufacturers are working on vertical axis designs. For reference, existing coal-fired power plants typically produce electricity at between 3¢ to 5¢/kwh.

Wind energy development in the western U.S. has been proposed by some to generate a substantial portion of increasing electricity demand. After all, the western U.S. does have its share of wind. Yet, it should also be obvious that the western U.S. is a long distance from the largest population centers and large electricity demand. For this to become effective, huge investments in infrastructure must be made. It is expected to take a decade or more just to obtain the permits and construct the required electrical transmission lines alone. Further, wind generation may be better considered to be simply energy generation rather than true capacity generation, since its generated power is only available when the wind is blowing. To maintain steady available power, a flexible peaking source should be expected to be required nearby. Without this separate peaking capability, transmission can easily exhibit frequency instability.

Solar is another promising energy source. China currently generates roughly 80% of its hot water from solar. Water can be heated to the point that any bacteria in the water are killed, and the water can thereby be made safe for drinking. Photovoltaic (PV) cells now convert solar energy into electricity at around 15% efficiency. In 2005, traditional crystalline silicon photovoltaic cells cost around $2.70 per installed watt of generating capacity, and it still has a relatively long payback period without some form of subsidization. When capital costs are considered, PV solar energy typically generates electricity at around 20 to 40¢/kwh, which is much higher than coal. But photovoltaic cells are expected to continue their rapid improvement in energy efficiency, and PV solar may rival the cost of electricity on the grid within five years. A company called First Solar claims it can already produce systems for $1.14 per installed watt. The company SunPower claims to have the best efficiency and lowest price per watt, even though their modules are more expensive. Another company, Suntech Power (out of China), is known for its quality, scale, and decreasing costs.

A second form of solar energy is being installed, which has been more prevalent for large utility installations. This form is known as thermal solar, and these solar plants collect solar energy to convert water into steam, sometimes utilizing an intermediate heat transfer fluid. There are three types of these thermal solar plants. First, parabolic mirrors can be utilized to direct the solar energy to tubes that carry a synthetic oil heat transfer fluid. The energy absorbed by the oil subsequently converts water into steam, which is then used to drive a steam turbine generator. There are currently nine of these large solar energy generating systems (SEGS) in the California Mojave desert. The initial unit was 14 MW and used parabolic through solar mirrors. Units 2 through 7 were 30 MW each, and the next two units, 8 and 9, were 80 MW each. A second means similarly uses molten salt in a solar tower. The heat is absorbed by the molten salt to convert water into steam and then drive a steam turbine. The third means uses small hydrogen engines on each solar collector that are driven by the solar energy.

One benefit of each of these solar designs, particularly the molten salt application, is that some of the generated energy can be stored for later distribution. This contrasts with wind generation, where the variability of wind generation is still a problem. Power has been generated from these solar collector units at about 6 to 8¢/kwh. A new solar collection system developed in 2008 by a firm called Skyfuels projects their new system price will be 4 to 6¢/kwh.

Waves and tides also offer a new frontier for generating electricity. This frontier is too new to provide cost or efficiency data at this writing, but it is an area with potential and will need to be monitored going forward.

There may be other renewable energy sources that can be expected to be developed in the future as well. However, if they are not being pursued today, these sources are unlikely to become mainstream anytime soon. Other non-renewable energy sources are covered next.

A Hydrogen Economy

There is much talk about hydrogen being a fuel of the future.[13] Either as a clean burning fuel or used in fuel cells, hydrogen could power cars, heat homes, generate electricity, and more. When pure hydrogen is combusted, it produces only water and no CO_2 or other pollutants. Hydrogen is particularly efficient when it is used as a fuel in PEM fuel cells. Significant government and private funding and research programs are in progress on using hydrogen. Yet, there are still enormous hurdles that must be overcome before a hydrogen economy becomes a reality.

The first hurdle is that hydrogen is so reactive that molecular hydrogen is not naturally found on Earth except in small quantities in the upper atmosphere. In nature, hydrogen is nearly always combined with other elements. Water and hydrocarbons are the common examples. Therefore, hydrogen cannot simply be gathered and then burned, which is unlike wood, coal, oil, or natural gas. No matter what source is chosen, a significant amount of energy must first be expended in order to release the hydrogen. In fact, most sources of hydrogen, like water, require even more energy to be expended than the energy obtained when the hydrogen is combusted—due predominantly to thermodynamic and chemical inefficiencies. This is a major problem. Hydrogen therefore cannot be considered a primary energy source. It is more appropriately considered an energy carrier—and methanol or even methane would have better properties for that. This is the primary reason that onboard vehicle hydrocarbon-to-hydrogen reformers are being pursued ahead of PEM fuel cells.

A second hurdle is that some of the physical properties of hydrogen make it unattractive. Hydrogen is the lightest element. This presents problems in transporting it, storing it, using it in gaseous form, or condensing it to liquid form. In fact, hydrogen atoms are so small they can ionize and easily diffuse through steel and other metals causing what is known as *hydrogen blistering* and *hydrogen embrittlement*, particularly at elevated pressures and temperatures. Hydrogen is also extremely flammable and explosive. Its flammable range is extremely wide, and it therefore has enormous safety issues. Hydrogen is currently generated in existing industrial facilities, particularly in oil refineries. However, these facilities build dedicated generation units and transportation systems to ensure the safety and security of supply. Constructing a hydrogen generation and supply system sufficient to be a main energy supply system would be expensive. A completely new supply system puts hydrogen at a further economic disadvantage, as opposed to natural gas or gasoline, which both have existing supply systems already in place.

A third hurdle for hydrogen is how it would be economically generated and supplied in the large quantities needed for widespread use. Of course, hydrogen is released from numerous chemical reactions and processes. But generating hydrogen for fuel in this manner would require these primary processes to be extensive enough to yield hydrogen in mass quantities for hydrogen to have the scale required to become a mainstream fuel. This appears unlikely—with the possible exception of hydrogen generation from widespread nuclear.

Hydrogen is currently produced by several means. Hydrogen is most economically obtained in quantity from natural gas (methane) via steam methane reformers (SMR). Currently, 96% of the global hydrogen production comes from fossil fuels, and roughly 50% is produced via SMR. One of the problems with this is that it generates large amounts of CO_2. Natural gas is the most common feedstock for SMR hydrogen production because of its cost, availability, and its high hydrogen to carbon ratio, which also minimizes its CO_2 emissions. In this process, natural gas reacts with steam over a nickel-based catalyst in reactor tubes at high temperature and pressure. The product mixture of CO and hydrogen then goes through a gas shift reaction using an iron-based catalyst, whereby the CO reacts with more steam to produce more hydrogen plus CO_2. The resulting hydrogen is usually purified by absorption techniques, and the CO_2 byproduct is predominantly vented to the atmosphere. The long-term, large-scale production of hydrogen from methane or other fossil fuels is not considered logical for a widespread hydrogen economy, as it would not solve the issues of efficiency or energy security, and it would be unnecessarily difficult and costly to capture and sequester the CO_2 byproduct.

Generating hydrogen from integrated gasification combined cycle (IGCC) coal technology may be the lowest, near-term cost alternative. A benefit of generating hydrogen by this means is that it allows the cogeneration of hydrogen along with electricity, and it therefore has a higher thermal efficiency. However, generating hydrogen by IGCC has disadvantages, such as generating more CO_2 than other means of producing hydrogen. This might be overcome with carbon capture and storage (CCS), but CCS would add considerable construction costs and would have higher operating costs relative to an IGCC without CCS. IGCC is also a less mature technology than other means of generating hydrogen. Overall, more environmentally sustainable methods of generating hydrogen may be needed than using coal for the source.

Hydrogen can also be obtained by electrolyzing water. Currently, 4% of the global hydrogen production comes from the electrolysis of water. Generating hydrogen by this method is used only when very high purity hydrogen is required. Electrolyzing water is roughly four times as expensive as hydrogen from SMR technology. Additionally, the best commercial electrolysis process today has only a 75% thermal efficiency. In other words, it requires about 130% of the energy to be input to the process, relative to the energy content available by combusting the hydrogen product. It appears unlikely that high-purity hydrogen generation by this simple method will become a prevalent fuel.

Gasification or pyrolysis coupled with steam reforming is another mechanism which could be utilized to produce hydrogen. Currently, however, such plants operate at only 26% efficiency, and hydrogen production costs are over $7 per Kg of hydrogen, which is about $62/mmbtu.[14] Hydrogen might also be produced in the future from biomass, yet this means of producing hydrogen is currently quite immature.

Hydrogen could someday be produced in concert with nuclear power. Since nuclear reactors do not emit CO_2, this solution might be particularly attractive. New generation nuclear designs are planned to operate at higher temperatures (1,300°F to 1,800°F versus 600°F to 750°F), and these designs would be better suited for high-temperature electrolysis. The thermo-chemical splitting of water can be achieved efficiently at 1,500°F to 1,800°F by using a chemical cycle such as the iodine-sulfur cycle. In this cycle, iodine and SO_2 are added to water where an exothermic reaction forms sulfuric acid and hydrogen iodide. At temperatures above 700°F, hydrogen iodide decomposes to hydrogen and iodine. At temperatures above 1,600°F, sulfuric acid decomposes to SO_2 and water and oxygen. The iodine and SO_2 are recycled in the process.

Nuclear Power

The nuclear option holds a large future promise, and the revival of nuclear plants appears to be imminent.[15] Nuclear energy might become limited due to fear, as people naturally fear what they don't understand—which applies to nuclear energy and radioactivity— but nuclear energy is now realizing an amazing swing in public attitude. Nuclear-powered plants do not combust fossil fuels or emit CO_2. Nuclear waste is now considered to be less threatening than it was thirty years ago. The U.S. licensing process has been greatly improved. The latest designs are known as generation III+, and there are currently only five pre-certified designs—such as Westinghouse AP1000—versus the hundred-plus designs of the past. These generation III+ designs are inherently safer relative to past designs.

Nuclear power currently accounts for almost 16% of global electricity generation and almost 20% of U.S. electricity generation.[16] There are currently 104 commercial nuclear reactors operating in the U.S. and about 440 worldwide (a handful are in the process of shutting down). Existing nuclear plants were designed and licensed for a forty-year life, but twenty-year license renewals are being granted with reasonably inexpensive plant modifications. The next surge in nuclear plant construction is expected to double the number of nuclear plants worldwide. The U.S. Nuclear Regulatory Commission (NRC) has recently received its first permit application since the 1970s and since Three Mile Island in 1979. These companies were racing for the 1.8¢/kwh production tax credit for innovative technology that was to expire at the end of 2008. Existing nuclear plants have typically generated electricity at around 3¢ to 4¢ per kwh, so this tax credit is quite significant. The NRC has received seventeen combined operating license applications, which represent twenty-six nuclear units. This new wave of new reactors will be considered within the next three years or so. New nuclear plants are being estimated to cost around $10 billion for a 2,000 MW capacity plant.

It will take ten to twelve years before any of these new plants could start up, due to the time it takes to permit, order equipment, and construct a plant—and this assumes no lengthy public opposition.[17] This timeline is extended partly because there is currently only one steel plant that can produce heavy nuclear forgings—Japan Steel Works. Over the last thirty years, there was just not enough nuclear work to keep all the past nuclear-forging-certified steel plants operating. This should change of course with an increase in demand, but it will take years to retool and recertify capable steel plants.

Many contend there are still some very tough issues that have not been satisfactorily addressed with nuclear power. One issue is how the waste will be handled. The U.S. national burial site at Yucca Mountain, Nevada, is still not functional, even after spending over $9 billion on it. Yucca Mountain may not be a long-term solution for nuclear waste, as it has only 65,000 metric tons of storage capacity and we already have approximately 55,000 metric tons needing to be stored.[18] Many nuclear power plants are currently storing their nuclear waste on site for now, expecting a future solution. One solution is with future nuclear designs, which are projected to be capable of burning the current nuclear waste. Yucca Mountain could then be utilized only for the really nasty military waste.

Another issue with nuclear energy is terrorism. There are many opinions on how severe this risk might be. Security in existing plants has increased significantly since 9/11, but there is still a concern and some terrorism risk. Future nuclear designs are expected to prohibit the ability to utilize their fuel for a nuclear weapon. But radioactive materials can still be a bit scary to some. There has always been a concern about the safety of operating these plants. But the general public is inherently poor at assessing risks such as these.

Finally, the world has about eighty-five years of known uranium reserves to feed nuclear reactors, and this is without much recent exploration.[19] But this does not account for another one hundred years of fuel from the current inventory of nuclear waste that could be burned in future reactor designs. New generation IV designs include pebble bed reactors having inherently-safe three-inch balls of fuel. Future designs could be either helium or sodium cooled. Fuel is also envisioned to become available from future breeder reactors. In a couple of decades, the estimated cost to produce electricity might very well be in the range of 2¢ per kwh. At some point in time, maybe around 2060, nuclear fusion could also become available. If low-cost nuclear power does transpire, nuclear energy could become *the* energy driver of the global economy within a few short decades and thereby initiate a sharp decline in the demand for coal to produce electricity. A sizable reduction in GHG emissions would therefore result.

Coal and Carbon Sequestration

Coal-fired power plants are the most dominant source of energy for global power generation.[20] Coal accounts for 50% of power generation in the U.S., 70% in India, and 80% in China. Pulverized coal (PC) power plants have typically generated electricity at between 3¢ to 5¢/kwh. Coal-fired plants are among the cheapest sources of power we have available to us today. Only hydroelectric power is (usually) cheaper. Worldwide coal deposits are also quite vast, and global demand for coal is expected to outpace every other energy source. The USDOE/NETL projects that by 2030 coal is expected to dominate the CO_2 production from power generation at 88%, with natural gas accounting for 10% of CO_2 emissions and oil accounting for only 2%. In addition, coal currently produces about 40% of the CO_2 emissions from *all* energy sources (power, industry, commercial, buildings, transportation, and residential). Since there is so much coal use, and it is projected to increase further, and coal generates such significant quantities of CO_2, coal-fired plants are prime candidates for carbon sequestration, otherwise known as carbon capture and storage (CCS).

Evaluations of CCS by the USDOE/NETL indicate the U.S. has about six hundred years of natural CO_2 storage sinks, and the world has thousands of years of CO_2 sinks. These sinks include deep geologic saline formations, depleted oil and gas fields, and coal seams. The most beneficial way to store CO_2 may be within oil and gas fields, as this can be economic for enhanced oil recovery (EOR). This is already being done at the rate of 2 BCF of CO_2/day in the U.S. CO_2 for EOR is most appropriate when partially depleted oil fields are close by. There are currently only four known EOR CCS projects in the world.

The DOE-NETL has the *goal* of helping develop CCS that

can capture 90% of CO_2 with 99% storage permanence at a 10% increase in the cost of energy. But this is only a goal at this point. The prospective technologies to allow this include molecular sieves, microporous metal oxide frameworks, and biological sequestration. The current outlook is much different, however, and the difference between the goal and the current situation defines a technology gap. With projected technology, the costs to operate the CCS plants could be from 14% more to a 40% increase in energy needed relative to a PC plant without CCS just to capture its flue gas CO_2.[21, 22] It appears the reason for this variance is that the 14% projection allocates the capital cost of the required system and amortizes it across the production from the plant, while the 40% projection takes the loss of generation of the plant into consideration. For example, an operating coal-fired power plant with a 500 MW output that then adds CCS to the plant would have a resulting output (available to sell) of only about 300 MW. Even with a lack of clarity on the technology, the operating costs for CCS have been estimated at about $20 to $30 per ton for capturing the CO_2, another $5 to $10 per ton of CO_2 for transportation, and storage costs could be $10 to $20 per ton of CO_2. Researchers at MIT have suggested that CCS costs might be able to get to $30 per ton of CO_2.

The construction cost for CCS is another big consideration. A March 2007 study at MIT suggests electricity generated from pulverized coal (PC) power plants with CCS would be 60% more expensive than from PC plants without CCS. The DOE has estimated CCS would currently add 70% to 100% to the price of electricity. IGCC is known as clean coal technology. But even electricity generated from coal-fired IGCC plants with CCS would be 35% more costly than IGCC plants without CCS. Note that Alstrom, a company who operates both PC plants and IGCC plants, states that IGCC plants alone (without CCS) are 10% more expensive than PC plants.

Significant breakthroughs in both technology and cost reduction will be required for coal-fired CCS to work well at reducing global

CO_2 emissions. It is expected that it might take ten to fifteen years for these hurdles to be overcome. Further, the quantity of CO_2 emissions needing to be sequestered will be high, and the transportation and storage of the CO_2 could require infrastructure of a similar scale to the current natural gas infrastructure. For these reasons, extensive CCS is not likely to occur for at least a couple of decades. Yet, even with these technical and economic challenges, CCS for coal-fired power generation is widely considered to be essential for any significant CO_2 emission reduction plan to be successful.

Another concern about CCS is not its increased energy requirement, its operating cost, or its construction cost, but the risk of environmental impacts and the potential loss of lives that could result from future Earthquakes or volcanic activity that might suddenly release the sequestered CO_2 from underground formations and into the atmosphere. It is not that this risk is considered high, but the risk and countermeasures dealing with the risk must be considered as part of the solution. Who has the legal liability, and for how long, of future risks such as these are issues that still need to be resolved with CCS.

A Summary of What We Can Do

To summarize, virtually everything presented here should be pursued to one degree or another. Some of the action measures show clear, early impacts on reducing GHG emissions. Some measures have energy efficiency and energy security benefits; plus, some of the measures should be economic and improve GDP. Measures that should be pursued include improved insulation, improved heating and cooling, improved lighting, improved CAFE standards, hybrid vehicles, more efficient engines, smaller cars, second-generation bio-fuels, cellulosic ethanol, sugarcane-based ethanol, bio-butanol, Jatropha oil, algae to fuel, forestation measures, more efficient electricity generation such as cogeneration, wind energy, solar energy, nuclear power, economic and marginal carbon capture and storage, smart and limited cap-and-trade policies, plus extensive research and development efforts on energy technologies such as oil shale, clean coal, and smarter natural gas usage. Each of these should indeed be pursued. But the intensity with which each of these are pursued might vary based on their potential. Some of these measures may prove to have little or no impact, and others will surely have more impact.

Unfortunately, we can't accurately foretell which technologies will win out, but some pretty good educated guesses can be made. Nuclear power may be our best hope at achieving scale. Wind will be limited geographically, but it is already relatively economic and could become even more significant. Natural gas should be plentiful for quite a while, and it has relatively low GHG emissions. Solar has a ways to go to be economic, but breakthroughs and improvements are likely with R&D. Algae are promising, but it still needs lots of R&D. Cellulosic ethanol has promise but might be limited in the feedstock required to achieve relative scale, and it will be dependent on multiple breakthroughs to become economic. Coal with CCS is also dependent on breakthroughs and geography, but coal reserves are very prevalent.

While each of these technologies are being pursued, it would be best to manage the near-term future by allowing and promoting domestic drilling and even allowing the processing of unconventional heavy crudes. However, just drilling anywhere is not smart or profitable. Many of the more probable locations for production are currently off limits. Increased domestic drilling would improve our energy security, help minimize huge trade imbalances, and should shore up our economies to help prevent economic collapse prior to the predominance of these other pursuits. While increased domestic crude production would be valuable, new U.S. refineries are not needed for the foreseeable future, absent occasional hurricane damage.

An Estimate of Probable Future GHG Emissions

The Stern Report directive is to reduce 60% to 80% of all greenhouse gas

STEVEN E. SONDERGARD

emissions by 2050. The IPCC action plan targets global GHG emission rates to be 50% to 85% of 2000 levels by 2050. The Lieberman-Warner bill mandates a 63% to 71% reduction in all U.S. greenhouse gas emissions by 2050. Yet are these targets reasonable? What is a reasonable expectation for both global and U.S. emissions by 2050? How successful are we likely to be at reducing CO_2-eq emissions?

To answer those questions, an estimating methodology is needed that could provide a reasonable answer even when many actual aspects of the problem are unknown. One such methodology is called a Fermi estimate, and it is attributed to Enrico Fermi who used it in his work on the Manhattan project, the project to develop the first atomic bomb, and in his teachings at the University of Chicago. Reasonable assumptions are first made about various aspects of the problem, and if any are in error, they are often offset by another assumption that is in error in the opposite direction. This type of estimating technique is often used for forecasting, and these foundational aspects will be used here to estimate probable future GHG emissions.

The reduction of GHG emissions could be realized in several ways. First, there could be incremental reductions in the combustion of fossil fuels via improvements in energy efficiency. Second, alternative energies could be used that emit reduced GHG emissions or in some cases virtually zero GHG emissions. Third, GHG emissions could be sequestered. Fourth, GDP or population growth could be decoupled from the production of GHG emissions. Fifth, population could be reduced or GDP could be sacrificed. The problem can be broken down simply by first estimating reasonable overall growth factors (by either GDP growth or population growth), estimating a decoupling factor of emissions growth to overall growth, and then picking reasonable improvement factors for each emission sector.

- Overall Growth by GDP—a reasonable assumption can be made about future GDP growth. GHG emissions have historically been strongly correlated to GDP, so this is a logical starting point. It is realized that GDP growth may not materialize, particularly if economic energy becomes unavailable. A global GDP growth of less than 2.5% per year is sometimes considered recessionary. IEA projects global GDP will increase at 3.6% per year to 2030. If we assume a conservative value of 3.0% per year global increase in GDP from 2005 to 2050, the global GDP would be roughly 3.78 times the level of 2005—$(1.030)^{45} = 3.78$. Without any mitigation measures, that translates into 3.78 times the 30 giga tons per year of GHG emissions today, or about 113 giga tons per year globally in 2050. Historical U.S. GDP growth has averaged about 2.5%. However, future U.S. GDP growth is likely to be less than that of the rest of the world, so 2.0% growth per year is assumed for the U.S. By 2050, the resulting U.S. GDP would be about 2.44 times that of 2005—$(1.020)^{45} = 2.44$.

- Overall Growth by Population—if GDP growth is not used, an assumption for population growth can be made. Forecasting equations for GHG emissions can be based on either GDP growth or population growth. If they are based on population growth instead of GDP growth, the average GDP per person is essentially assumed to remain the same, which is not considered to be as accurate. IEA projects the world population will be 8.2 billion by 2030. Almost all of the +1.8 billion increase is expected from developing countries (+1.7 billion). If the 6.4 billion in global population (2005) continues to grow at the projected exponential rate, the population in 2050 will be between 9.5 billion and 12 billion people. That will mean the increased CO_2 emissions from human respiration alone might be between 2.8 and 3.6 giga tons per year in 2050. If we assume a population of ten billion people in 2050, the growth factor would be 10/6.4 = 1.56. Because the highest GHG emission impact is expected to come from developing countries (due to a combined population increase and an increasing GDP per person), it is believed

STEVEN E. SONDERGARD

that GDP growth would have a better correlation with GHG emissions growth. The subsequent estimating equations will therefore use GDP growth instead of population growth.

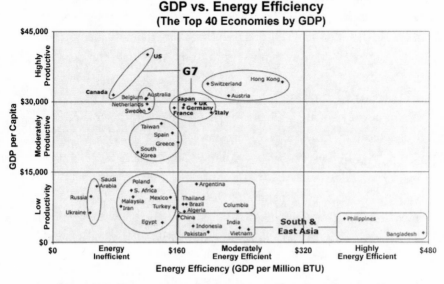

Figure 13.7. GDP vs. energy efficiency. (Adapted from: Peter Corless with permission)

To determine overall emission growth factors, it is helpful to understand some comparative relationships of what has already been achieved. Refer to Figure 13.7, which is a plot of energy efficiency versus GDP per capita. Notice that the results vary widely. The North American countries have a highly productive GDP per capita, but they are relatively energy inefficient in terms of GDP per million British Thermal Unit (BTU). The average G7 country is only slightly better in energy efficiency. But it is Hong Kong, Switzerland, and Austria that appear to each have both high GDP per capita along with energy efficiencies that are roughly twice that of the North American countries and nearly 30% better than the other G7 countries. These are indicators of potential. Based on this data, it is assumed that the global GDP-to-GHG correlation can be decoupled by 30%. Note that this decoupling is in addition to the reduction assumptions presented in the subsequent paragraphs for each emissions sector. Since the U.S. has one of the highest levels of GHG emissions per GDP of any country, it is assumed the U.S. can decouple its GDP-to-GHG relationship more than other countries, and by a painful 50%, which is significantly more than the expectation for the rest of the world.

The Stern Report and the USEPA breakdowns will be used as the framework for the global and the U.S. emission sectors respectively. Recall that the Stern Report states the global GHG emissions in the year 2000 break down as follows: power sources (24%), land use (18%), industry (14%), transportation (14%), agriculture (14%), buildings (8%), waste generation (3%), and other (5%). The USEPA states the U.S. emissions in 2005 break down as follows: residential (17%), agricultural (9%), transportation (28%), commercial buildings (17%), and industry (28%).

- Power sector—this segment was about 24% of global GHG emissions in 2005. The U.S. GHG emissions from all industry was about 28% in 2005, and it is estimated that two-thirds of that percentage was from the power sector, or 18.7%. There are several emission improvement measures that could be implemented within the power sector. Nuclear power does not generate CO_2 emissions, but it does have other issues. Cogeneration is also a substantially better alternative, effectively doubling the energy efficiency of the traditional power generation process. Renewable power such as hydroelectric, geothermal, solar, wind, and bio-energy could also help reduce emissions, but their increased use is not expected to decrease the power sector emissions by appreciable amounts relative to the size of the sector. Pulverized-coal-fired boilers dominate this sector, and they are expected to continue their worldwide

STEVEN E. SONDERGARD

domination. The main hope is that CCS can significantly reduce CO_2 emissions from these plants. Fuel switching from coal to natural gas could help reduce near-term emissions, but this action in effect simply transfers the problem to future generations if we do not develop adequate low-emission energy sources such that our coal reserves do not get used eventually.

- For the GHG emissions calculation, it is optimistically assumed that between 2005 and 2050 nuclear power generation (having zero GHG emissions) will constitute 30% of all *new* power generation for both the world and the U.S. It is also assumed CCS will be utilized on 40% of all global fossil fuel power generation (new or existing) and 60% of all U.S. fossil fuel power generation. Further, the applied CCS will be able to recover 70% of all GHG emissions by 2050 from the hydrocarbon-fired power generation. Additionally, it is assumed that future CCS will use only 20% more energy to perform this recovery. Each of these assumptions is aggressive.

- Lighting—of the total worldwide power generated, 19% goes toward lighting. This portion of the power sector should be able to radically improve its GHG efficiency, as it is expected that we could improve the efficiency of lighting systems by a factor of about seven. Fluorescent lights use about 25% of the electricity as incandescent lights, but the latest LED lights use only about 15% of the wattage for the same lumens. Technology will probably make further improvements by 2050, but conversely, 100% global conversion cannot be expected either. On the other hand, restricting the purchase of inefficient light sources can be predominantly mandated through regulation. Overall, it is thought that an optimistic position on lighting would be an assumption that we will be able to reduce the power sector emissions due to lighting to 15% of today's levels by 2050.

- Land/forests—this sector was 18% of global GHG impact in 2005 and is due to the declining ability to assimilate CO_2 through depleting forests. Reforestation, reduced deforestation, afforestation, and using forest products for bio-energy are all possible. The IPCC WG3 states 65% of the total mitigation potential for this sector is located in the tropics. Alas, the probability of altering this decline is not considered very high due to the disparate countries involved and their strong drive for their own growth and development. An optimistic probability of 30% improvement is therefore assumed by 2050.

- Industry—this sector was 14% of global GHG emissions in 2005. The U.S. industrial sector was about 28% of U.S. GHG emissions in 2005, but this includes the U.S. power sector. Currently the U.S. industry sector generates 2.25 giga tons per year of emissions. Coal-fired power generation constitutes two-thirds of the U.S. industry emissions (therefore 9.3% of overall emissions, or 1.5 giga tons of GHG per year), and it is assumed non-power-sector industry can improve efficiencies by 30% prior to 2050 relative to 2005, both for the U.S. and globally. U.S. industry has been on a steady track of improvement now for over ten years. This improvement is partially due to energy improvement projects and partially due to industry moving overseas where reduced emission controls exist. These two drivers somewhat offset each other in global emissions because the overseas regulations where these plants relocate are not as strict. The current improvement measures include better burners for fired heaters and boilers, high-efficiency motors, flue gas energy recovery, low-level heat recovery, more efficient air systems, heat integration, and reduced emissions in general. Global measures are expected to be similar, but they might actually track a bit behind the U.S.

- Transportation—this sector was about 28% of U.S. GHG emissions and about 14% of global GHG emissions in 2005. The improvement measures include improved gasoline engine fuel efficiency, improved diesel fuel efficiency, exhaust emission controls and capture, hybrid vehicles, public transportation, and urban planning. It is assumed that vehicle fuel efficiency could be essentially

STEVEN E. SONDERGARD

doubled and that transportation would have 50% of the GHG emissions in 2050 relative to 2005.

- Buildings—this sector was about 8% of global GHG emissions in 2005. The improvement measures include insulation, improved glazing efficiency, improved lighting efficiency, using sunlight, passive and active solar, using smart thermostats, improved efficiencies in air conditioners, improved overall heating and cooling efficiencies, and improved kitchen appliance efficiencies. The 2007 IPCC WG-3 projects that roughly 30% of the GHG emissions from buildings can be avoided by 2030. The U.S. building sector is further broken down into commercial and residential. The commercial segment was about 17% of U.S. GHG emissions in 2005. This segment is assumed to have a 50% improvement in emissions efficiency in 2050 relative to 2005. The residential segment was also about 17% of U.S. GHG emissions in 2005. This segment will have more difficulty improving than the commercial segment, due to the hurdles of capital, poverty, sheer number of residences, building design, and enforcement difficulties. The magnitude of these barriers is higher in undeveloped countries. This segment is assumed to have a 30% improvement in emissions efficiency in 2050 relative to 2005. These building improvement assumptions are over and above improvements in lighting, which is included in the power sector estimation factor.
- Agriculture—this sector was about 14% of global GHG emissions and 9% of U.S. GHG emissions in 2005. The improvement measures include improved carbon storage in soil, restoration of lands, improved rice cultivation measures, livestock manure management to reduce CH_4, nitrogen fertilizer application to reduce N_2O, and improved energy efficiencies. This sector is assumed to have a 50% improvement in emissions efficiency relative to 2005.
- Waste and Other—these segments were a combined 8% of global GHG emissions in 2005. The improvement measures include waste minimization, recycling, landfill methane

recovery, waste incineration with energy recovery, composting, and controlled waste water treatment. It is assumed that an optimistic 50% improvement in global emissions relative to 2005 could be achieved in the waste and other sectors.

The Estimating Equation for *Global* Emissions in 2050 Using a GDP Multiplier

$$30 * 3.78*70\% * (24\%*(60\%+40\%*30\%*120\%)*(81\%+19\%*15\%)*(84\%*70\%+1)/1.84$$

growth   power   no CCS   hydrocarbon with CCS   lighting impact   new nuclear power plants [**]

$$+ 18\%*70\% + 14\%*70\% + 14\%*50\% + 14\%*50\% + 8\%*50\% + 3\%*50\% + 5\%*50\%)$$

land   industry   transportation   agriculture   buildings   waste   other

= 45.5 GT/yr

note [**] The stipulation for 30% of *new* power plants to be nuclear results in an extra algebraic term (for the 84% increase in world electricity and 34% increase in U.S. electricity from these calculations).

The result from this estimate tells us that even with optimistic technology improvements, the global GHG emissions might be about 50% *greater* in 2050 relative to 2005. This is hardly the 60% to 80% *reduction* that the Stern Report targets for 2050. Of course, this answer is based on somewhat debatable guesses for improvement. We could pick a multitude of improvement guesses and basically come up with almost any answer we choose. However, if the equation is used as a tool and better and better educated guesses are used in the equation, it becomes increasingly clear that reducing the global GHG emissions at all in the future will be a hugely daunting task. For example, if a more aggressive growth forecast is taken, with the global GDP averaging a 3.5% increase in emissions per year (which is what it has actually averaged in recent years), the multiplier to

    STEVEN E. SONDERGARD

2050 becomes 4.70 rather than 3.78, and the answer becomes 56.5 GT/yr, which is 88% greater than the global emissions of 2005.

The Estimating Equation for *U.S.* Emissions in 2050 Using a GDP Multiplier

8.0*2.44*50%*(18.7%*(40%+60%*30%*120%)*(81%+19%*15%)*(34%*70%+1)/1.34

growth   power   no CCS   hydrocarbon with CCS   lighting impact   new nuclear power plants **

+ 9.3%*70% + 28%*50% + 17%*50% + 17%*70% + 9%*50%) = **5.55 GT/yr**

remaining industry   transportation   commercial   residential   agriculture

The result from this U.S. estimate indicates that with optimistic GDP growth, technology improvements, and further decoupling of emissions to GDP, the yearly GHG emissions from the U.S. might see a reduction of about 30% by 2050 relative to 2005. Although this falls a bit short of the 60% to 80% reduction the Stern Report targets and the 63% to 71% reduction the Lieberman-Warner bill targets, it does provide a little hope (albeit with considerable hardship and pain). If a more aggressive growth forecast is taken and the U.S. GDP averages 2.5% increase per year (which is what it has averaged in recent years), the multiplier to 2050 becomes 3.04 rather than 2.44, and the answer becomes 6.91 GT/yr, which is only 14% less than the emissions of 2005.

These calculations indicate just how large a factor GDP growth, or alternatively, human population growth, will be to the emissions produced in the future. The projected correlation between GDP, or alternatively population, and GHG emissions is recognized as quite important to future GHG emissions. If the GDP or population were to plateau or possibly be reduced, or if the correlation between GDP and GHG emissions were significantly decoupled, the hope for reducing future global GHG emissions becomes much higher. Further, restricting future population growth has other obvious environmental and societal benefits. Controlling population would automatically control a very large lever of GHG emissions. It would also reduce the strain on food, water, and many other natural resources. A coordinated policy for population control may be unlikely, but it appears to be a measure that should be pursued with some vigor, and probably with even more intensity than climate change itself, if we desire to extend our society's survival.

What provides considerable hope for improving GHG emissions from the U.S. is that it currently has the highest GHG emissions per GDP of any country, and it is thought that this high level could be reduced to that of other industrialized countries without a hugely negative GDP or security impact. A significant amount of this hope lies with the ability to achieve improvements in emission reduction technology and with carbon capture and storage. It would not be wise to bet against man's ingenuity when his back is against a wall. But success by this means *does* imply that man's back must indeed be against a wall, and that requirement alone will have societal implications.

The preceding calculations indicate it will be difficult to stop the current trend of escalating emission rates, and it will be even more difficult to *reduce* GHG emission rates—even *with* technology improvements—without a catastrophic societal collapse. If this analysis is reasonably accurate, then our best track appears to be a path that would not jeopardize our society unnecessarily on more than one front by going too far in our zeal to stop Global Warming but instead to take measures that fall short of that jeopardy. A wise leader should assess whether any battle can be won before entering into it, and he should assess the outcomes and costs of the battle. The action measures required to limit GHG emissions and the trade-offs with our economies can be considered such a battle. Pursuing action measures that do not have negative economic impact or that would have a limited negative economic impact appear to be the best approach toward the problem rather than radical high-stakes alternatives that risk our society. Minimal impact action measures probably include many of the measures portrayed in the Vattenfall diagram (Figure 13.5). Additionally, the setting of policy, providing the proper incentives, and aligning the driving forces to allow the best possible outcome should accordingly be established, while keeping in mind there are other societal threats acting upon us.

Alternate Growth Perspectives

ExxonMobil, 2007, has a well-researched view of the future global energy picture.[23] Although this is obviously an oil company perspective, an accurate appraisal of the energy picture is critical to their future, and it is not too dissimilar from alternate analyses. From 2005 to 2030, a projected global GDP growth of 3.0% per year is forecasted, and most of this will be from developing countries. The energy intensity, however, is expected to improve, and the actual world energy demand is expected to grow at only 1.3% per year average as a result. The yearly growth breakdown by sector is transportation at 1.7%, power generation at 1.5%, industrial at 1.2%, and residential and commercial both at 0.7% per year. The energy growth from 2005 to 2030 is therefore expected to be $(1.013)^{25} = 1.38$. With this growth, the overall global energy efficiency and emission improvements must be at least 28% just to avoid *any* increase of GHG emissions by 2030 [the calculation is $(1.38–1.00)/1.38$]. A 28% reduction in emission efficiency is quite significant.

Using the emissions reduction measures projected by ExxonMobil, the global growth rate of energy related CO_2 emissions is forecasted to be 1.2% per year for all sectors. Actually, the non-energy related CO_2 emissions are forecasted to be higher per year than the energy related CO_2 emissions. If the best case is assumed of 1.2% emissions growth per year for all sectors, then a 35% global increase in CO_2-eq emissions from 2005 to 2030 might be expected. The equation for this is $(1.012)^{25} = 1.35$. (Note: in late 2008, ExxonMobil updated its energy outlook to 2030. In this update, the global energy demand was reduced from 1.3% per year growth to 1.2% per year growth. The forecasted global GDP growth remained the same at

3.0% per year, but the energy intensity improved to 1.7% per year. The forecasted global energy growth became +29% in 2030, the biofuels+wind+solar energy growth became +9.3% per year, and the forecasted global CO_2 emissions became +28% in 2030 relative to 2005.)

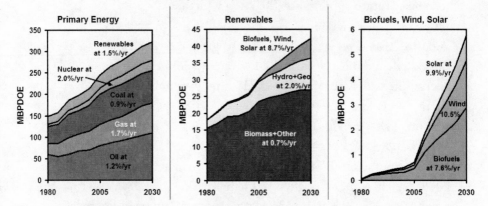

Figures 13.8 a, b, and c. World energy supply projections. Note that renewable energy has rapid growth, but it is still dwarfed by fossil fuel consumption (coal, gas, and oil). (Adapted from: ExxonMobil, Energy Outlook, November 2007)

For another comparison, the IEA, 2007, projected a 57% increase in CO_2 emissions by 2030.[24] This is after the effect of global mitigation measures. Again, in late 2008, the IEA updated its world energy outlook. The IEA now forecasts the energy demand will increase 1.6% per year to 2030, and the global CO_2 emissions will increase a cumulative 45% by 2030.

Whichever analysis we choose to use for the CO_2 emissions increase—either the +35% by 2030 (per ExxonMobil, 2007) or the +29% by 2030 (per ExxonMobil, 2008), or the +57% by 2030 (per IEA, 2007) or the +45% by 2030 (per IEA, 2008), or even the +50%

STEVEN E. SONDERGARD

by 2050, calculated by the estimating equations above, the prospects of *reducing* GHG emissions do not appear to be too good before mid-century. From Figure 13.8, note that even though renewable energy production is increasing rapidly, it should not be counted on to save the day by 2030 and perhaps beyond. The magnitude of global fossil fuel use is simply too immense, and it will not be easy for renewable fuels to approach the scale of the global energy demand that has been the result of available and inexpensive energy for over a century. Even if scale can eventually be achieved with renewable fuels, it will likely take decades and have enormous costs.

A real benefit of having estimating equations as presented above is to perform sensitivity analysis on specific variables in order to see how they might change future GHG emissions and to determine what it would take to reach certain GHG emission targets. A sensible cost/benefit study based on cause and effect could then be performed to help steer policy and action. These further calculations, however, are beyond the scope of this work.

14. SHOULD YOU BELIEVE EVERYTHING YOU READ?

Focus: A sampling of inaccurate and extremist positions and reporting from both sides of the climate change argument are presented.

Layman's Brief:

Political and media battles have been waged concerning climate change that could be called ferocious in nature. Games have been played to win these battles that claim the science is settled or claim a certain high number of scientists are behind a specific stance. Neither of these claims mean the truth is being represented well. There has obviously been much written on the subject of Global Warming, yet the disclosure is often selective and the message frequently errant (intentionally or otherwise). There are many scientists that allow their paradigms or emotions to sway their perspective about Global Warming and the ramifications of climate change. People in general, including scientists, are naturally inclined to believe the prevalent stream of thought, particularly when it benefits their livelihood or might be outside their narrow field of expertise. Moreover, most of what is relayed about climate change is directed by political bodies and the media, not by scientists.

Consequently, of the two opposing camps, the camp believing in severe anthropogenic Global Warming is the more prominent. Proponents of this view tend to be much better funded and tend to receive much better press. Those who tend to question the severity of Global Warming are usually more silent, perhaps due to their concern about public and private ridicule, having their funds cut, or even losing their jobs. The objective evaluation of opposing views has been dangerously limited because of this dichotomy and subsequent misrepresentation.

Quick Reference:

- The issue of acting on Global Warming is deeply rooted with more and more people.
- However, it appears that very few people have examined the stepwise cause-and-effect relationships and many are flying blind in their beliefs. Then there are those who, for some reason, find it easier to manipulate the facts to fit their beliefs or to promote their own stance rather than to be open to a change in their viewpoint.
- Some past reports of the significance of global warming show an altered graph of past average surface temperatures that basically eliminated both the Medieval Warm Period and the Little Ice Age altogether. This graph is now known as the hockey stick. Its construction is attributed to climatologist Michael Mann et al. from their work in 1998, and it was prominently featured in the 2001 IPCC TAR, *An Inconvenient Truth*, the Stern Report, *Beyond Oil and Gas: The Methanol Economy*, the McKinsey Climate Change study, and many other works. This hockey stick graph is now predominantly dismissed as misleading, particularly after a congressional investigation and the

National Academy of Science graciously reported it as such in 2006. The graph was *not* included in the 2007 IPCC AR4 report for this reason. Yet, this fabricated depiction is still widely used and referenced.

- Al Gore admitted in a May 2006 *Grist* interview that overrepresentation of the facts is an appropriate tactic in order to promote Global Warming. Even though his film has been highly popular and earned Al Gore great accolades, *An Inconvenient Truth* isn't impartial about its message. Even scientists who support its underlying message must agree that this film is shallow in its scientific content, unbalanced in its presentation, and perhaps irresponsible in its one-sided representation of data. On the other hand, it should be a lesson in just how scary the situation can be to those without a balanced view.

- Phillip Cooney, a Bush political appointee, named White House Council on Environmental Quality from 2001 to 2005, has been accused of editing at least three federal climate change reports "to align these communications with the administration's stated policy." Mostly, he repeatedly replaced the word "will" with the word "may" and inserted the word "potentially" in statements about the role of human activity in climate change, effectively watering down the reports. He also redacted much of a paragraph from one report, which is what Al Gore highlighted in his film.

- Dr. Christopher Landsea, a leading expert on hurricanes and tropical storms at the National Oceanic and Atmospheric Administration (NOAA), resigned as an author of the 2007 IPCC report due to politicizing of the data: his research, and the research of others, showed there was no global warming signal found in the hurricane record, although a lead author for the IPCC, Dr. Kevin Trenberth, reported the polar opposite. Actually, NOAA stated in April 2007 that each of their eighteen global climate models consistently predict *reduced* hurricane activity from global warming, due to a robust increase in vertical wind shear in the tropical Atlantic and the eastern Pacific Oceans. Although it would be statistically possible for a severe hurricane to escape the expected increase in vertical wind shear, increased hurricane activity might not be expected to be the norm.

- Al Gore's graph on the monetary damages from past U.S. hurricanes and flooding shows a soaring trend, even before Hurricane Katrina. Yet, this graph uses actual dollars spent, which are not adjusted for inflation, increased population growth near the coasts, or more expensive structures built near the coasts. With these adjustments, the graph looks entirely different, showing no upward trend.

- Two authors have stated their work was misrepresented in the 2001 IPCC summary, including atmospheric physicist and MIT professor of meteorology, Dr. Richard S. Lindzen.

- Professor Lindzen has more recently stated, "scientists who dissent from the alarmism have seen their funds disappear, their work derided, and themselves labeled as industry stooges. Consequently, lies about climate change gain credence even when they fly in the face of the science" (April 2007).[1]

- Willie Soon and Sallie Baliunas, 2003, declared the reality of the Medieval Warm Period and that it was warmer than today.[2] However, the article used a reportedly errant assumption that any changes in past precipitation indicated a rise in temperature. This article is now forwarded as employing faulty methodology, and its publication resulted in the resignation of half of the periodical's editorial staff.

- Dr. Frederick Seitz, past president of the National Academy of Science (NAS), was dominantly ignored with regard to the 1995 IPCC report. He stated in a *Wall Street Journal* article that this IPCC report is a "disturbing corruption of the peer-review process" and "it is not the version approved by the contributing scientists listed on the title page." Dr. Seitz continued, "the IPCC reports are often called the consensus view" but "whatever the intent was of those who made these significant changes, their effect is to deceive policymakers and the public into believing

STEVEN E. SONDERGARD

the scientific evidence shows human activities are causing global warming."

- Even with these noted concerns, it should conversely be noted that the IPCC is a consortium of top scientists that generally have rigor in their data gathering, data consolidation, and reporting. The IPCC updates their climate change analysis every few years, and their work is generally considered to be among the best overall scientific assessments of the subject. But the IPCC is far from infallible, and human emotions and previous stances have been known to override the objectivity of opposing scientific data, particularly when the final reports are written by government representatives.
- It appears that political drive, political momentum, or political correctness have driven scientific censorship to either prohibit speaking out or to diminish an opposing viewpoint.
- These occurrences of purposeful censorship from both sides are very disturbing. Our actions should be supported by the facts on this potentially high-impact issue. The consequences of our actions, if based on anything else, could be unnecessarily severe.

More Detail:

The examples cited here are some of the more widely referenced and are only representative of the problem. Climate change is indeed complicated. Those that report on the subject often slant their descriptions and do not convey much, if any, information opposed to their viewpoint. For this reason, many fail to carefully take the extra time to gather and weigh the arguments from both sides of the subject. In addition, for one reason or another, people generally start with an unsupported conclusion on the subject and then shore up their conclusion with one-sided facts. This does not represent intellectual maturity. Idealistic stances are not appropriate on this matter. Yet, that is exactly the stance most are taking. Governments are not helping either—even though they have a duty to act with more reason. But this is not unusual. "Mankind, it seems, makes a poorer performance of government than almost any other human activity. In this sphere, wisdom, which may be defined as the exercise of judgment acting on experience, common sense and available information, is less operative and more frustrated than it should be."[3] If you are reading this far, you are already armed with much more balanced information than the great majority have obtained. Now you, the reader, have somewhat of an obligation to help increase intellectual maturity and help overcome the polarization between idealistic stances.

The Michael Mann Hockey Stick Graph

The graph now known as the MBH98 *hockey stick* is attributed to climatologists M. E. Mann, R. S. Bradley, and M. K. Hughes from their work in 1998[4]; and together with a pre-1400 extension by Mann in 1999 and a spliced instrumental temperature series, the resulting graph was prominently featured in the 2001 IPCC TAR, *An Inconvenient Truth*, the Stern Report, *Beyond Oil and Gas: The Methanol Economy*, the McKinsey Climate Change study, and many other documents and publications. It looks detailed and convincing.

This fabricated graph has been widely publicized, and it is still extensively used and referenced often. Unfortunately, it is now predominantly dismissed as misleading, particularly after the National Academy of Science (NAS) graciously reported it as such in 2006. The NAS tried to be conciliatory by stating in their summary that the Earth is warming without precedent—at least over the last four centuries. Previous conclusions utilizing the original MBH98 graph stated that late twentieth-century temperatures were the highest within the last 1,000 years and late twentieth-century temperatures are unprecedented. Yet, it is now known that these conclusions were made by handling the data poorly and are simply unwarranted. "The dataset used to make this [graph] contained collation errors,

STEVEN E. SONDERGARD

unjustified truncation or extrapolation of source data, obsolete data, incorrect principal component calculations, geographical mislocations, and other serious defects"—all of which substantially affect the temperature index.[5] Consequently, the graph was *not* included in the latest IPCC AR4, 2007, report.

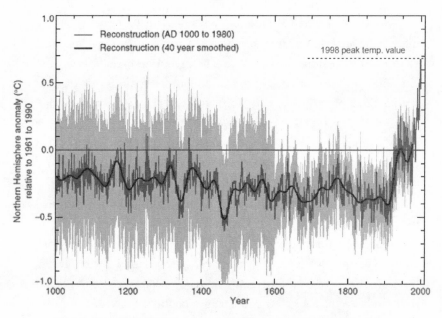

Figure 14.1. This MBH98 figure is also known as the hockey stick. Note that the error bands are relatively wide and that the Medieval Warm Period and the Little Ice Age are not indicated on the figure, plus the 1998 global average temperature appears to be unprecedented. (Adapted from: *Nature*, 1998, v01.392, pp.779–787)

A reconstructed graph prepared from the same source data yet with substantially improved quality control has since been attempted utilizing the same MBH proxy methodology. This work was first performed by Canadian Stephen McIntyre and economist Ross McKitrick. They handled the data with greater care, and their corresponding graph yields a Northern Hemisphere temperature index that indicates twentieth-century temperatures do not stand out relative to the preceding centuries in either their high values or in their variability. Additionally, "the errors and defects in the [original] MBH98 data means that the indexes computed from it are unreliable and cannot be used for comparisons between the current climate and that of past centuries."[6]

This conflict and fiasco actually resulted in Congress requesting two investigative reports, which led to congressional hearings to get to the bottom of the issue. The preeminent statistician, Dr. Edward Wegman, was called to lead this investigation for Congress with a panel of statisticians to help with the task. It became readily apparent the graph was inherently difficult to produce over this time period. Remember, thermometers were not available until Galileo invented them in the late sixteenth century. It took more than a couple of centuries after that for anything more than very sparse temperature records to be created. Early temperature measurements for this graph were therefore obtained by proxy data. The proxy data includes tree ring widths, coral growth, pollen counts, and ice core records. There were 112 proxy series used for this graph, but only fifty-five of them went back to the sixteenth century, and only twelve went back to the fourteenth century. How all this proxy data was handled was of utmost importance to the conclusion. The treatment of the proxy data, particularly with the sparse nature of the proxy data, is complicated. The statistical methods proved more crucial to the results than the climatology itself. The entire handle of the hockey stick is the result of a modeling process from proxy temperature data, while the recent trend is from weather station data. A hockey-stick-shaped graph would be the normal expectation from such a process, no matter what the data.[7]

In the end, Wegman's panel discredited the work of Mann et al. and supported the work of McIntyre and McKitrick. Wegman further noted that less than stellar statistical handling was not

STEVEN E. SONDERGARD

uncommon in climatology and atmospheric science research. He also unveiled serious problems with peer reviews and the lack of objectivity within the tight circles of the paleoclimate community.

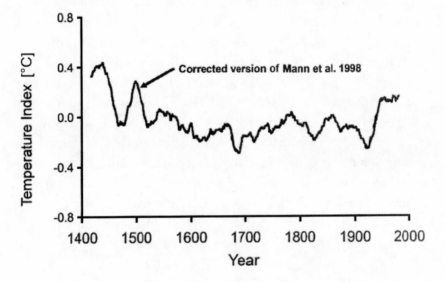

Figure 14.2. The corrected version of MBH98. Note that this corrected version indicates that the Medieval Warm Period was warmer than today but that the Earth is warmer today than the temperature averages during the preceding four centuries. (Adapted from: Energy & Environment, vol. 14, No. 6, 2003, p.766, figure 8)

Even after this congressional hearing concluded, the battle continues today. Michael Mann has since reworked his proxy temperature measurements in an attempt to support his conclusions. Ross McKitrick has found further problems with the proxy database, and he continues to debunk Mann's work. The Web site at http://www.realclimate.org and the Web site at http://www.climateaudit.org are representative of both sides of the argument.

Phillip Cooney Editing

Phillip Cooney was a Bush political appointee and former oil industry lobbyist who was named Chief of Staff for the White House Council on Environmental Quality from 2001 to 2005. He has been accused of editing at least three federal climate change reports. His stated objective was "to align these communications with the administration's stated policy."[8] Cooney made 181 changes in the administration's 2003 Strategic Plan for Climate Change Science Program. Many of these changes were preserved in the final document. Most of the editing involved the repeated replacement of the word "will" with the word "may," and he inserted the word "potentially" in statements about the role of human activity in climate change, effectively watering down the reports. Cooney also inserted references to possible benefits from climate change. He further redacted much of a paragraph from one report, which Al Gore highlighted in his movie. Cooney claims he based his edits on NAS documents.

Dr. Kevin Trenberth's Claim about Increased Hurricane Activity

Dr. Kevin Trenberth is head of the Climate Analysis Section at NCAR and a lead author for the IPCC. Dr. Christopher Landsea is a leading expert on Atlantic hurricanes and tropical storms at NOAA (National Oceanic and Atmospheric Administration). Landsea was invited by Trenberth to serve as an IPCC author as well. In the aftermath of Hurricanes Katrina and Rita, a press conference on Atlantic hurricanes was scheduled at Harvard Medical School's Center for Health in October 2004. Landsea voiced his concern about the scripted message, and within days of the press conference, Landsea was replaced by Trenberth, even though Landsea was the specialist. Statements were made by Trenberth at the press conference that were alarmist in nature, and they caught widespread media attention. The statement was broadly relayed by the press that experts were warning that global warming would likely continue to spur more outbreaks of intense hurricane activity. Yet, Landsea's own research did not substantiate this claim, nor did he know of any other research that substantiated the claim.

Landsea's response to Kevin Trenberth was, "there are no known scientific studies that show a conclusive physical link between global warming and observed hurricane frequency and intensity."[9] Landsea then brought the matter to the IPCC, who had reported in 1995 and 2001 that there was "no global warming signal found in the hurricane record."[10] Landsea asked the IPCC, "where is the science ... what studies are being alluded to ... the IPCC process taints the credibility of climate-change science." The IPCC took the position that the press conference was not an IPCC event and Trenberth had the right to give his opinion. The IPCC's chairman, Dr. R. K. Pachauri, responded to Landsea that Trenberth "did not in any way misrepresent the IPCC and apparently his statements accurately reflected IPCC's TAR."[11]

Actually, NOAA has recently (April 2007) stated that each of their eighteen global climate models consistently predict *reduced* hurricane activity from global warming, due to a robust increase in vertical wind shear in the tropical Atlantic and the eastern Pacific Oceans. Although it would be statistically possible for a severe hurricane to escape the expected increase in vertical wind shear, increased hurricane activity might not be expected to be the norm. Landsea later resigned as an author of the 2007 IPCC report due to gross politicizing of the data, and he stated the IPCC itself was compromised. The latest IPCC AR4 now downplays a correlation between hurricanes and global warming.

Al Gore's Hurricane/Flooding Cost Graph

Al Gore's graph on the monetary damages from past U.S. hurricanes and flooding shows a soaring trend, even before Hurricane Katrina. This graph uses actual dollars spent that are not adjusted for inflation, population growth near the coasts, or more expensive structures built near the coasts. Even after Katrina, reconstruction is being pursued within very high risk areas. It is realized that these areas have been the home to families for generations, but this just doesn't seem very smart. Some people will naturally accept more risk, and there will always be those who push the envelope. Others facing the requirement for change simply refuse to adapt. Government simply cannot protect all of these people from their own poor decisions and unwitting behavior. Nevertheless, there *are* some effective and economic action measures that make sense versus other high-cost, delayed-impact global warming countermeasures. These action measures include levies and dykes, constructing protective sea walls, restricting development in the lowest areas, elevating building foundations, and using better building techniques.

As previously stated, a predicted increase in hurricane activity from global warming is not supported by any of the eighteen climate models at NOAA. There is also a historic thirty-year cycle in hurricane activity that went completely unmentioned in Gore's work. The statement that hurricane intensity will increase the costs of damage within the U.S. is almost exclusively the result of increased construction in harm's way and the ongoing general escalation of construction costs. Hurricane damage and coastal flooding will impact more people primarily because more people are moving to coastal, flooding-prone areas. After adjusting for inflation, population growth near the coasts, and increased construction near the coast, the graph looks entirely different, showing no upward trend. The worst decade for both total and major hurricanes was the 1940s, followed by the 1950s, the 1930s, and then the 1890s. The 1990s were not abnormal in this regard. The current decade is obviously not yet over to conclude anything.

The IPCC Statements about Human Impact on Warming

The recent IPCC statements about anthropogenic Global Warming are very carefully worded. The lead authors are government representatives that essentially wrote the final reports with sometimes selected input from their scientific teams. The usefulness of the legion of scientists was in effect to provide credibility by simply having their names listed among the long register of expert authors. This gives the illusion of active involvement and indicates high consensus, even when their scientific input is often neglected.

Let's examine a few of the statements in the IPCC reports. The 2001 IPCC report (TAR) stated a "66% to 90% chance that human activities are driving recent warming." Taking some liberty with interpretation, all that this statement really claims is that there is a reasonable confidence level that human activities are more than 50% of the cause of any observed global warming. It does not state that human activities are the *only* cause of recent warming, but the careful IPCC wording could lead the reader to draw that conclusion.

The 2007 IPCC report (AR4) states "most of the observed increase in global average temperatures since the mid twentieth century is very likely due to the observed increase in anthropogenic greenhouse gas concentrations." In addition, the AR4 states "it is very likely that global climate change is not due to known natural causes alone." Again taking the same liberty in interpretation, these statements say that mankind has very likely caused at least *some* of the observed global warming and probably *most* of the observed global warming, although confidence levels are not included. The assessments of this author are not in complete disagreement with the essence of those IPCC statements. However, there is still a fair amount of uncertainty that is not addressed or reported by the IPCC, particularly around solar variability, solar feedback mechanisms, and cosmic ray impacts on low-elevation cloud formation.

The IPCC report goes on to state that anthropogenic warming and rising sea levels are projected to continue for centuries even with the stabilization of greenhouse gas emissions. The facts do appear to indicate there is a feedback delay of the current CO_2-eq concentration due to the Earth's inability to assimilate the full quantity of current emissions. However, the IPCC fails to mention that rising sea levels are expected to occur naturally. No matter what we do to reduce anthropogenic emissions, this expectation is indicated by past interglacial cycles and melting glacial ice. In addition, rising surface temperatures would also be consistent with the data surrounding past interglacial cycles, which were all natural in their cause.

The carefully worded IPCC statements might mislead some because the wording itself creates a greater impact with a reader than the actual content of the underlying statement. It can also be observed that several key aspects concerning the statements are frequently left unmentioned and thereby could also mislead the reader.

Further, the overwhelming weight of future warming forecasts by the IPCC is primarily based on AOGCM modeling, which was shown earlier to be at the very least questionable. Finally, as was also pointed out earlier, the magnitude of the forecasted warming by the IPCC using their AOGCMs appears to be higher than other available information would suggest. Because of these arguments, the integrity and honesty of the widely-trusted IPCC reports have been questioned by more than a few scientists.

15. WHY IS THERE SO MUCH CONTROVERSY AND EMOTION ON THE SUBJECT OF GLOBAL WARMING AND CLIMATE CHANGE?

Focus: Survey results of global warming are first examined. Observations are made about agendas, driving forces, and human nature. This leads to some answers, but they are only speculative in nature.

Layman's Brief:

There are many reasons why people believe or communicate what they do. One common reason appears to be that people reinforce or convey ideas that support their personal paradigms. Another reason is that any given belief might be held simply because it is idealistic in nature. Beliefs of this sort have an elegant appeal to many. Others are more malicious in that they advocate partial truths or exaggeration in order to support a certain agenda. Others still are proponents of whatever makes their life easier or would cause the least personal disruption. None of these reasons change the real truth of course; they just distort the truth and impact what people believe is the truth. Unfortunately, these reasons can be ultimately destructive to society as a whole.

With climate change readily in the stream of consciousness of most of the literate world, any catastrophic climate event that occurs can be expected to be subsequently followed with a question about whether the event was a result of Global Warming. The question adds to the hysteria of the moment, and any slightly affirmative answer, whether contrived or not, supports those whose paradigm is aligned with that response. The press normally obliges, of course, because it makes a great story for the media.

But for those who are less hysterically driven, observing this process is quite frustrating. These individuals ask why the press must always resort to exaggeration and why more people do not take a more measured approach, with a judicious balancing of the facts resulting in reasoned conclusions.

The survey results presented here are scientific polls of American beliefs. The author's observations presented here are based on personal perception and are only speculative in nature. These observations are not based on scientific study, and they could easily be contested. They are forwarded so the reader can give them some consideration in light of what is already known.

Quick Reference:

- With all of the global warming information being communicated today, as much as 73% of Americans in 2007 called global warming a "very serious problem," with over half of Americans believing immediate attention is required. The tipping point for action, whether valid, right, or wrong, appears at hand.

- Many have jumped on the bandwagon for Global Warming simply because it is a cause that seems significant and makes us feel good. Yet, we should ask if we just want to *feel* good or if we actually want to *do* good.
- Some of the most fervent Global Warming advocates appear to have the strong underlying agenda of anti-industry, anti-capitalism, or anti-U.S. Some actually do desire an idealistic collapse of modern society as it exists today, and they do not consider or care about the individual consequences.
- Some who promote strong global warming actions desire to elevate the authority of man at the expense of God. Whether man should take action on global warming may depend on whether you consider God the central force in your paradigm of the universe or whether man must always take action to control his own destiny.
- Research funding and publication are paramount to most scholars. Global warming has therefore become big business. Researchers can be driven by greed, fear, and power to further their careers.
- Political leaders can easily embrace the fight against Global Warming because it is popular and it aligns their platforms with the noble cause of saving the planet.
- The media can be generally observed to create and perpetuate most any hysteria.

More Detail:

Survey Results about Global Warming

Many surveys over the last few years convey that Americans in particular are aware of global warming, acknowledge its reality, perceive a consensus, believe it is serious, and that America should do as much or more than other countries to address it.[1] Less than one-quarter of the public doubts the reality and significance of global warming. Yet, MIT completed a survey in March 2005 indicating Americans poorly understand climate change. A June 2005 PIPA poll indicated a slight majority (52%) perceives there to be consensus in the scientific community. In a September 2005 ABC/*Washington Post* poll, a slight majority of Americans (56%) are at least "mostly convinced" that there is global warming or a greenhouse effect is actually happening. A June 2005 ABC/*Washington Post* poll indicated a majority (61%) believe human activities are either "the most important cause" or "one of several important causes" of global warming. In a March 2005 Gallup poll, the majority (64%) do not believe the media is exaggerating the seriousness of global warming. A June 2004 ABC/*Washington Post* poll indicated two-thirds of Americans do not think global warming will pose a serious threat to their way of life in their lifetime. But 79% thought that global warming will pose a serious threat to future generations. A majority rejects the argument that taking action is too economically onerous and is optimistic that reducing greenhouse gas emissions will in the long run actually benefit the economy by increasing efficiency. In addition, the majority endorsing action is divided on whether the problem is pressing enough to take urgent high-cost actions or whether a moderate response is sufficient. If most other countries were to participate in the same action measures, support for taking action then becomes overwhelming with the large majority wanting to do as much as other industrialized countries to reduce emissions.

The Pew Research Center did a political survey in January 2007 on what Americans believe about global warming.[2] Curiously, the results vary quite significantly depending on political affiliation.

STEVEN E. SONDERGARD

Table 15.1

SURVEY RESULTS	Is our earth getting warmer?	Is it a very serious problem?	Needing immediate attention?
Liberal Democrats	92%	73%	81%
Moderate Democrats	83%	52%	61%
Independents	78%	46%	58%
Lib. & Mod. Republicans	78%	35%	51%
Conservative Republicans	54%	18%	22%
Total	77%	38%	55%

(source: The Pew Research Center, http://people-press.org/reports/display.php3?reportid=303)

President Bush also mentioned in his 2007 State of the Union speech that global climate change is a serious challenge. The politicians and the general public appear ready to take action on climate change. However, it appears they do not really comprehend what that implies or will require, and they do not fully appreciate the pain involved in taking high-impact action measures. Human nature also indicates we humans will be very reluctant to give up the creature comforts and conveniences that have become familiar. Therefore, the resolve to carry out any plan having high-impact action measures can be reasonably questioned.

An Observation about Significance

It appears many have jumped on the bandwagon for Global Warming simply because it is a cause that seems significant. This is actually predictable and laudable human behavior. It gives people purpose. It makes people feel good. This cause can provide the opportunity to save the world. Our world is increasingly complicated and overwhelming, and it is increasingly difficult for people to feel significant. Global warming is also a particularly complicated and multifaceted cause for which seemingly very few have taken the time to research and understand with appreciable depth. But it also has a simple, concise aspect that can be easily grasped. People follow the cause in order to be part of a solution that is bigger than themselves in order to feel significant, in order to do something good, in order to save the world. Those motives can all be considered good.

Taking action to reverse man-made greenhouse gases is obviously gaining momentum. But there may be other more important issues impacting our civilization. Some of these potentially more important societal issues might be the increasing population, global economic recession, restricted crude oil supply, water and sanitation, HIV/AIDS, malnutrition, and diseases like malaria. Further, climate change is a cause in which there currently appears to be somewhat hidden and dire probable downside(s) from overreaction. Caution is therefore proposed to guide our efforts, and intellectual maturity is warranted to remain nimble enough to embrace the best tactics as better information becomes available in the future. It is proposed that we at least be flexible and open to potentially more important issues that might present themselves that could even require a more focused effort than global warming. As Bjorn Lomborg puts it, "do we just want to do something that makes us *feel* good, or do we actually want to do something that *does* good?"[3]

An Observation about Society

Some of the most fervent Global Warming advocates appear to have the strong underlying agenda of anti-industry, anti-capitalism, or anti-U.S. Many that consider industry bad are either idealists that do not recognize the good industry provides society or they do not believe the good outweighs the bad. Many that consider capitalism bad may consciously or unconsciously lean toward egalitarianism or communism, and they do not realize the innovation that capitalism spurs is more than a zero-sum game. Nor do they appreciate the technological benefits the free-market engine provides to drive the leading countries of the world. Some may be simply so disgusted with the observed inequities in the system that they want it purged, even if the inequities are in the minority and even if civilization were to be threatened by doing so. Each of these advocates actually do desire an idealistic collapse of modern society as it exists today, and they do not consider or care about the consequences. Unfortunately, it appears that climate change is being used by groups having these underlying agendas as a pawn for their selfish or more subversive motives. Many that consider the U.S. bad believe the U.S. has acted with arrogance, or they simply believe the U.S. has too much power in the world. Without taking sides one way or the other, it is apparent the world does not appreciate the U.S. when it acts like a bully, and the global majority does not appreciate the U.S. acting as the world's police force.

An Observation of Religion

Interestingly, it appears that some who promote strong Global Warming actions desire to elevate the authority of man at the expense of God. This can be either conscious or unconscious. Assuming one has a healthy concern for people and society and one has some understanding of the issues, risks, and consequences—it appears that a strong contribution or possibly even core aspect determining one's position on whether man should significantly intervene against global warming boils down to an answer to the question of "Where is your hope?" Is your hope in God, or is your hope in the sustained longevity of man, with man controlling his own path and outcome? Another way to ask the question even more simply: Is man the central force in your paradigm of the universe, or is God the central force in your paradigm of the universe? If man is considered your central force, then man is the one who must take the action to provide the best outcome. This foundational issue seems to help explain why the argument can sometimes get so heated.

An Observation on Scientific Research and Publication

It is quite apparent that global warming has become big business and has resulted in billions of dollars flowing into scientific research and to scientists who would otherwise be without funding and perhaps purpose. This fact alone has the potential to sway many to be a part of the ride. There is little driving force for researchers to halt this windfall. University professors, for one, generally have tenure, and they basically cannot be fired, even if they were to slant the truth, fail to report the whole truth, or worse. It is research funding and publication that are paramount to most scholars. The examination of basic human nature tells us that people have pretty simple driving forces, and greed is at the top of the list. Scientists can indeed be swayed with money. The oil industry has been suspected of this for years, rightly or wrongly. But scientists can also be swayed by public opinion, particularly on aspects that are outside of their narrow field of expertise. Plus, the fear of reproach from one's peers is a consideration. Many summary reports appear to be *very* carefully worded for these reasons. Researchers would indeed need to have a strong moral compass, little concern about reproach, and an overriding goal of problem resolution and dissemination in order to overcome these natural drivers of greed, fear, and power.

It is also not uncommon for scholars to have expertise in only a narrow sliver of the entire subject. Connecting the dots and objectively weighing conflicting information without emotion require a different skill set altogether. Indeed, without a larger multi-faceted perspective, narrower views do indicate a much different picture. Whatever the reason, the possibility exists for reporting unrepresentative truth. The problem does not need to be the result of conscious activity or be widespread. But the potential for even the unconscious actions of a few could be significant enough to reach a tipping point. Readers should simply be aware of this possibility and be aware that these factors could be slanting your perspective. There are many good scholars that are now taking the extra steps to weigh opposing data and objectively question more details. The best approach for any scholar or reader is to not be afraid of opposing viewpoints and to glean valid information from multiple resources. Gather the facts from several sources and seek out opposing sides of the argument, study the issues, question everything, look at the facts objectively, and utilize intellectual maturity. Then decide for yourself.

Other Considerations

Political leaders can easily embrace the fight against Global Warming because it is popular and aligns their platforms with the noble cause of saving the planet. It allows these leaders to be the heroes of the moment, with the future costs and headaches becoming some future leader's problem. Sensible cost-benefit analysis can become quite secondary. Government organizations have a natural bent to push Global Warming because it is an easy scapegoat for the present condition and an avenue to the future. It not only provides a common foe, but it could rally constituents, enhance political power, redistribute wealth, and divert blame away from political ineptness. Political criticism could be reduced, and approval ratings could be considerably increased. A common foe can unquestionably unify political support. History indicates the foe does not need to be valid or real. Manipulation of a crisis would not be uncommon. Power is why many political leaders get into politics in the first place. The redistribution of wealth has long been sought by many who are dissatisfied with the present state. Pointing a finger at any new threat can also make it easier to proceed with alternate agendas when the action measures of the threat fit with these other agendas. Further, increasing taxes are palatable to many more constituents when they are also viewed as being good for the planet.

The media can be generally observed to lean toward creating hysteria around most any topic in order to both validate their existence and to sell their product. This effect can indeed generate emotional responses, at least until common sense is allowed to percolate over emotion. But this process can take years or sometimes even decades. It is not uncommon in human nature for anyone to accept data that supports one's own paradigm of reality and dismiss glaring data to the contrary. This fact alone can delay or prevent a balanced, objective evaluation from occurring. It is a small step from there for people to allow an end to justify the means. Emotional responses are then common when facts or circumstances threaten foundational beliefs. Emotional responses can be valuable when quick action is required, but they are rarely the best driver for balanced decision making.

The current ground swell of common perception is now so great that many government jobs, environmentally-oriented companies, research funding, non-government organizations, economic stimulus, and egos currently depend on the problem to remain, and it can be expected that these individuals and organizations will act to protect their interests, even when data indicates otherwise. This is not a good excuse, but it is human nature. In addition, the self-perpetuation of some groups dictates the requirement of a cause, and Global Warming currently fits the bill.

With these observations, it is not too hard to understand the initiation and perpetuation of our present condition.

16. A QUICK SUMMARIZATION— WHAT IS THE MOST LIKELY REALITY ABOUT GLOBAL WARMING AND CLIMATE CHANGE?

Focus: Fifteen realistic summary points are made about climate change, which are believed to be based on a well balanced examination of the subject matter. In short, it should not be too surprising that a realistic conclusion from the balance of facts is that some of the observed global warming can be attributed to increased anthropogenic greenhouse gases and some of the observed warming can be attributed to natural causes.

Many people ask for a simple summary of the science behind global warming. Even though many aspects are still very much unknown, a short summary is indeed possible. Yet, a summary will undoubtedly be controversial; it will most assuredly require modification as better information becomes available, and it may be somewhat dangerous. People may read only the summary with a closed mind about the subject, being near one polar extreme or the other on the issue, and may not look further at the breadth of facts, particularly if a summary is not aligned with one's initial belief. However, because others have and will offer summaries with limited impartiality, a summary is provided—with a little reservation. Here are fifteen summary points based on the data.

- Water vapor is by far the most significant greenhouse gas on Earth, and it is approximately an order of magnitude more significant than CO_2 (or any other gas). But water vapor is predominantly not anthropogenic (caused by humans) nor easily modeled, especially over the long periods required to determine climate trends.

- The climate of Earth has not been steady or constant, and it should not be expected to be steady or constant in the future. The climate of Earth goes through significant natural cycles from natural causes, and these cycles have larger temperature excursions than the present and recent trends of global warming over the last century.

- The acknowledged 1.1°F–1.3°F increase in the average global surface temperature over the last one hundred years appears to be partly anthropogenic and partly caused by natural cycles and natural phenomena, such as increased solar irradiance.

- Humans have likely caused about a 100 ppm increase in atmospheric CO_2 levels, and this increase has probably contributed about 0.6°F to as much as 0.9°F in the average global surface temperature during the last century. This includes all feedback mechanisms. Other anthropogenic greenhouse gases have caused an estimated 0.1°F rise in the average global surface temperature in the last century.

- The total solar effect, including the described solar feedback mechanisms that amplify the total solar irradiance, appears to have realistically caused maybe one-fourth to possibly one-half of the average surface temperature increase of the last century. Therefore, the total solar effect is estimated to have caused an increase of about 0.3°F to 0.6°F in average global surface temperature in the last century. However, some information suggests the impact from the total solar effect may be even greater. If this solar impact is confirmed and the total solar effect is more extensive than acknowledged, the contribution to the global temperature rise due to atmospheric CO_2 levels must, by necessity, be less.

- Warming at the poles appears to be two to ten times greater than the average global temperature warming. Winter warming is about twice that of summer warming. Nighttime temperatures are rising faster than daytime temperatures.

- The Earth's natural feedback mechanisms appear to currently have a multiplying effect that is less than double and probably somewhere around 1.7 on the direct radiative forcing impact of greenhouse gases. This statement still has considerable uncertainty, as Earth's natural feedback mechanisms are quite complex and are not yet well understood.

- Any small remainder of noted average surface temperature increase appears to be due to imprecise modeling, poor statistical handling of surface temperature data, and correctional uncertainties with available satellite data; that is, any small remainder probably doesn't really exist.

- Significant near-term effects from anthropogenic Global Warming appear unlikely until the atmospheric levels reach three to four times pre-industrial levels. But that does not mean there won't be any effects at all from anthropogenic Global Warming or that there won't be effects, potentially even significant, from naturally-caused climate changes.

- The possibility of a climate change or a climate jump in the future is more likely to be caused by changing solar output and other natural causes, as has occurred in the past, than increases in anthropogenic greenhouse gases.

- There is a strong possibility that our world's slide into the next ice age has already been delayed as a result of anthropogenic impacts on our climate, and this delay could potentially extend for thousands of years.

- The targeted stabilization thresholds of 450 ppm or 550 ppm for atmospheric CO_2 or CO_2-equivalent concentrations appear to be too restrictive for the good of mankind.
- The probability of both industrialized countries and developing countries to join in a coordinated effort to aggressively manage climate change by not exceeding the targeted 450 ppm or 550 ppm thresholds are seen as extremely small.
- There appears to be much more potential for significant human pain and suffering from human overreaction to Global Warming than from Global Warming itself. This point would lead us to embrace a plan of no-harm action and even low-impact action as most prudent—and also politically achievable—worldwide. But even this action still allows for quite significant improvements in our emissions of anthropogenic greenhouse gases and for delaying the impacts of forecasted outcomes.
- The global issues of rapidly increasing population, including strain on the world's food, water, and other resources, in addition to the impending decline in crude oil production, both appear to be potentially larger, more certain, and more near-term issues for mankind than do the impacts from anthropogenic climate change.

17. SUGGESTED POLICY

There are many alternative proposals for future climate change policy. In general, these might include either waiting for most nations to join in action measures or risk leading the pack, either allowing or promoting regional regulations or coordinating national and global regulations, either being aggressive or being conservative in regulations, either waiting for the economy to improve or stimulating both causes simultaneously, enacting a cap-and-trade mechanism to limit carbon emissions, driving hard for energy efficiency measures, promoting renewable energy, encouraging lower-impact fossil fuels, and funding research and development. Some of these are reasonably well thought out and even have merit. Others, however, are ideologically rigid, simplistic, problematic, or risky in nature. The merit of some options will lie in the scope and the detail. It is believed a careful balance of risk and reward would work best. The following bullets are suggested policy points that are supported by the aspects discussed within this book.

- There is much about our climate that is still poorly understood. The scientific details of global warming and climate change are far from settled, and it is misleading to declare a consensus on the matter.

- Global warming over the last century has been observed and quantified. The impact of anthropogenic greenhouse gases on global warming appears to be roughly half of what is commonly conveyed by the IPCC. In other words, it appears to be at the lower extreme of the probability range stated by the IPCC.
- There are other societal issues that appear to be at least as threatening as Global Warming. Yet, taking selected action measures to mitigate Global Warming would still be prudent.
- Several climate change technologies, practices, and policies have been targeted for implementation. They should be given high consideration if they move society toward energy conservation, energy security, reduced trade deficits, and economic stabilization. They should be questioned if they do not.
- The economies of OECD countries will probably suffer in the next few years, and economic stability will need to have elevated priority in relation to other societal issues.
- If the prices of crude oil and natural gas drop too far, consumers could get complacent again about energy conservation. This might be a good time for government to impose an energy tax of some kind or adopt a limited cap-and-trade policy. But policies should take care to consider their economic impact and avoid burdensome economic consequences.
- There are many known no-harm and low-impact action measures to counter Global Warming that have additional societal benefits, and they should be pursued with haste.
- The economic and societal impacts of the higher-cost mitigation measures could be relatively severe. Implementing many of these measures today does not appear to wisely allocate our human and natural resources.
- Currently-economic energy-saving mitigation measures are significant and will likely require mandating in order for most individuals to comply.
- A good portion of the potential mitigation measures could be effectively driven by economic business drivers in the

form of a carbon tax or limited cap-and-trade system, unless they go too far and create economic calamity.

- The pursuit of potential longer-term currently-uneconomic mitigation measures will require government assistance, particularly for research and development. Research on these measures should be supported by governments in order to have the best chance at achieving the radical technology breakthroughs that will be required for both controlling Global Warming and to achieve post-fossil-fuel energy sources.

- The global political cooperation required to fully mitigate Global Warming appears to be highly unlikely. But political cooperation should still be pursued in an attempt to successfully navigate the issues.

- China and India combined are expected to have almost 50% of the demand growth for fossil fuels and commodities in the next decade or two. These countries need to be involved for any policy to be successful.

- An uncoordinated patchwork of national, regional, and local efforts is expected to have detrimental impacts overall.

- Political action without dominant worldwide cooperation is expected to often be counterproductive to the cause.

- Even with full global political cooperation, a unified global effort to significantly reduce CO_2 levels would still probably have a low impact on the climate.

- Population, water, food, and energy are poised to become the largest concerns for society. The trade-offs of any action with these concerns should be fully considered.

- We should establish a good U.S. national energy policy that might include the following:
  - Provide a stable, long-term, sensible regulatory framework that would effectively guide business planning decisions and allow businesses to plan ahead and act.
  - Understand that we need to pursue most of our options for energy, in parallel. The sheer scale of our energy consumption will require it.

- Provide mechanisms that allow common sense to rule over partisan politics.
- Promote mechanisms that reduce our dependency on imported oil from unfriendly countries.
- Promote both near-term fossil fuel replacement and other longer-term energy solutions that are not fossil fuel based.
- Require more stringent energy conservation measures for both end consumers and businesses, without punishing the leaders.
- Set higher CAFE standards, encourage hybrids, and help develop newer engine technologies.
- Support for most first-generation bio-fuels should only be given in order to build infrastructure for next-generation bio-fuels.
- Champion next-generation bio-fuels.
- Cease the requirement for mandatory increases of corn-based ethanol.
- Eliminate the import tax on the more efficient ethanol made from sugarcane.
- Intensify the pursuit of domestic energy sources to increase energy security, reduce trade deficits, and avoid supply shortfalls.
- Support domestic unconventional natural gas production at scale.
- Promote some fuel switching to lower-emission natural gas to take advantage of prevalent North American natural gas reserves.
- Prepare for the crude oil supply/demand cushion to be threatened within a decade or less.
- Allow drilling in the Outer Continental Shelf, Arctic Ocean, and ANWR, plus support enhanced oil recovery to delay a pending oil crisis by one to two decades. This might be enough time for alternative energy technologies to be developed. Even partial pursuits would yield partial results.
- Do not promote the need for new U.S. refineries but

STEVEN E. SONDERGARD

support and assist with flexible conversion units for diesel and distillate production.

- Properly support nuclear power to become a more mainstream generator of new electricity demand in just over a decade.
- Support R&D for economic CCS, principally for coal, as it is expected to remain the dominant world power generation fuel for decades to come, particularly in developing countries.
- Support research and development on inherently safe and flexible nuclear reactors, battery technology, algae to fuel, cellulosic ethanol, wind, solar, IGCC, and carbon sequestration.

18. FINAL SUMMARIZATION

We live in an interesting time, with progressively more complex social issues. It is therefore increasingly difficult to determine what our top societal priorities should be. The combined state of expertise concerning global warming and climate change might be described as a relatively early state in which we don't fully know what we don't know, and we have considerable uncertainty with what we do know. The Global Warming problem could indeed become a top societal threat, but there is ample evidence suggesting other societal issues will be at least as severe. Detractors of Global Warming have portrayed it as junk science without statistical robustness or a good foundation. There do appear to be multiple instances where the statistical and scientific rigor can be questioned. But that does not mean there is no merit to every aspect of the Global Warming arguments. Consciously or unconsciously, most who report on the science behind global warming magnify the supporting points of their own paradigms and ignore opposing points. A balanced, realistic view does not deny that anthropogenic Global Warming is occurring. But it also suggests Global Warming may be less grave than what is portrayed by the IPCC and others.

Further, the commonly targeted emission levels and reduction requirements of Global Warming proponents appear to be unnecessarily restrictive, and the measures required to remain under the targeted emission levels appear to have a high likelihood of having real detrimental economic impacts for society. Even if civilization could meet these targeted emission levels, many of the forecasted impacts can be expected to occur anyway. The prescribed requirement to break the creature-comfort habits of humans will be very hard and appears unlikely without global mandates. Research and development on new technologies will be required on several fronts. Coordinated policy, such as a carbon tax or cap and trade, is expected to be pursued. A global political system doesn't yet exist to coordinate policy and allow realistic accomplishment of targeted requirements, particularly within developing countries. Complete global political coordination is often assumed, but it is considered naïvely optimistic.

A dozen years from today, it is projected that Global Warming will not be viewed as having as much of a concern as is often relayed today. There are other pressing social issues, such as a rising human population, a projected shortfall of crude oil production, and economic security that appear to be greater societal threats than anthropogenic Global Warming, and these appear to be more of a near-term threat to modern civilization. Increasing population and the lack of inexpensive energy could lead to shortages in water and food or lead to economic instability and even war. Addressing the societal threat of the increasing global population would reduce anthropogenic emission impacts, prolong many scarce resources, and reduce numerous stresses on our civilization and our globe. In addition, there is little argument against conserving fossil fuels, and it should become a societal priority for many reasons. Energy conservation would have the benefits of reducing emissions, prolonging scarce resources, and increasing energy security, particularly for the U.S. and other oil-importing countries. Economic stability is a recent threat that could eventually prove to be extensive. Without address-

STEVEN E. SONDERGARD

ing these threats, it appears our civilization is likely headed for societal collapse. If we are not successful at overcoming these threats, the risks of famine, military war, and economic collapse are quite high in this century. Even before these forecasted threats are upon us, there could be painful trade-offs between food, water, housing, transportation, oil, electricity, and many commodities including steel. What should be obvious is that any progress that can be made with controlling population and conserving fossil fuels would not be inconsistent with the mitigation of greenhouse gas emissions.

Elevating the priority of the alternative threats does not mean climate change itself is not an issue. Nor does it mean there might not be negative consequences to climate change. What it means is that our combined and focused attention should be directed with pinpoint accuracy at the specific causes that have the largest probable impact and nearest-term impact to society. It seems logical that our efforts should be focused on the pressing societal issues that we have a reasonable opportunity to remedy. Even though the issues of population growth, energy security, economic security, and climate change are greatly intertwined, we may not be quite hitting the nail on the head if we focus only on climate change. Future generations depend on our clearest thinking here. There is still substantial scientific uncertainty concerning global warming. Some of the causes of global warming appear to be out of our control, and the probability of successfully mitigating global warming appears to be quite low. Fighting climate change too aggressively appears to have significantly greater downsides than most acknowledge, and it has the potential to be very counterproductive. The economic and societal costs of fighting the wrong foe would not only be inefficient, it could be disastrous given limited human and natural resources.

If everyone knew the real risks and if they had a choice between the two risks of global economic collapse or global climate change, what would they choose? Would people rather change their habits, avoid the risk of a climate catastrophe, and accept a higher risk of economic collapse? Or would people rather retain their habits, accept the risk of climate change, and defer the risk of economic collapse? I submit that the majority would choose the latter, particularly when one's own well being is in jeopardy. In fact, history tells us most people will struggle in any effort to change their habits even when they want to.

Any fight against climate change will be like fighting an adversary that has never been fought before. This book should help add some clarity and perspective concerning the benefit of the fight, the probability of winning the fight, and assessing some of the costs of the fight. Not only will the fight require a higher coordinated effort than mankind has ever been able to achieve in the past, but mankind itself is targeted as the adversary. Probability analysis of future scenarios as well as comparing upside probabilities with downside probabilities of this battle against climate change appears to tell us that it may not be in our best interest to fight this war with too much over-zealousness. Our society will be under increased stress from many fronts in the future. There are multiple issues pressing upon society that could end in collapse. Many of these are admirably listed in chapter sixteen of *Collapse* by Jared Diamond. This book lists the most serious environmental threats as being destruction of our natural habitats, declining wild fisheries, losing genetic diversity, eroding agricultural soils, limited fossil fuels, depleting underground aquifers, limited world photosynthetic capacity, increasing unnatural toxic chemicals, the transfer of species to non-native areas, increasing greenhouse gases which increase warming, growing global population, and the generation of waste materials. Which one(s) of these will collapse our society? We don't really know for sure. But Jared Diamond does provide ample evidence that multiple past human civilizations have collapsed due to one or more of these causes. We should learn from the past and be aware that our modern society does not appear immune.

With all of these societal pressures, in order to provide ourselves the longest-term security, it does not make sense to add a high level of economic stress and possible military war on top of it all. If we destroy the very thing we are trying to save, is it worth it? It appears much more prudent to steer our civilization toward an approach that has a greater probability of success and thereby pursue a climate change action plan that falls short of societal upheaval and a plan that embraces no-harm action and low-impact action. This is especially true if an all-out pursuit would have only a small chance of meeting prescribed targets for sustained CO_2-eq levels and if pursuing low-impact actions would yield a similar environmental result. To be clear, no-harm actions are those that are already economic, and these should all be diligently pursued. Low-impact actions probably include most of the mitigation measures shown in the Vattenfall figure shown earlier. Many of these measures could have an economic payout with a higher price of oil, and some of the measures would conserve energy too.

Completing both no-harm actions and low-impact actions probably correlates to reductions in global GHG emissions by about 20% to *maybe* as much as 30%. High-impact actions *might* be able to increase these improvements by another 20% or so. Yet GHG targets are being discussed and might soon be set at a 60% to 80% reduction. This does not appear to be realistic. It should be realized that reaching aggressive targets such as these would only be feasible with the development of multiple, radical breakthroughs in technology *and* by halting both population growth and virtually all the combustion of fossil fuels globally. If the breakthroughs are not realized, and probably even if they are realized, the societal pain of reaching these targets would be immense. If aggressive GHG targets are indeed set, mechanisms to monitor the progress of technological breakthroughs and progressively alter policy according to societal pressures would be prudent.

The future does appear to hold increased difficulties and turbulence. But these problems can be solved. The pitfalls are apparent, if we objectively look out for them, and they could be avoided. What we have before us is something resembling a multi-dimensional minefield, where we have to maneuver through this minefield to safely move civilization forward. The mines are actually encroaching upon us as our population increases and natural resources diminish. We cannot continue on the same course, and we cannot stand still. A measured approach in selected directions through time will be required. We should not ignore climate change issues or we risk longer-term consequences. We cannot move only to support climate change initiatives or we risk other societal issues overtaking us. We cannot only support renewable energies or we risk near-term energy security and economic inefficiency. We cannot only support fossil fuels or we risk future energy security and technology shortfalls. We cannot rely too extensively on ethanol made from corn or we risk limiting our food supply and the cost of that food supply. Mandated objectives alone would meet with resistance. Regulations alone would be insufficient. We must pursue energy efficiency measures, but that should only be the start. We should pursue measures that are economic, but economic measures alone will not prepare us for post-fossil-fuel energy. We should pursue government-backed R&D, but R&D alone would be deficient. A carbon tax or cap-and-trade policy could help steer the course of action, but aggressive policy could have dire economic impact. We should pursue energy conservation measures until they become uneconomic at "high" oil prices. We should pursue the use of more natural gas, as it appears to be both abundant and has low GHG emissions. We should selectively pursue heavy oil production (e.g. cleaner oil sands) or risk oil supply shortfalls. We should pursue next-generation nuclear energy or risk needed scale. We cannot afford to sit still, and we should not move too aggressively. A balanced approach will be required to be successful.

Much more open-mindedness and clarity will be needed when examining the subjects of global warming and climate change to avoid the pitfalls. One of the main problems has been the two camps that have extreme positions on the subject, and neither perspective listens well to an opposing view or attempts to work through conflicting issues. There are valid points on both sides, and there are nonsensical arguments from both positions. The attempt was made in this book to unveil some of the complex dynamics surrounding climate change. This work has made an attempt at a balanced review, and it reflects as much of the truth as the author has been able to objectively research up to the point in time of this writing. It is hoped that the contrasting perspectives presented here will provide the opportunity for a little more intellectual maturity to be exercised.

It is realized that the desire for a realistic and balanced stance as presented by this book may not be accepted by many. This book may be uncomforting to most people who have existing views on the climate change argument, as it does not firmly support either extreme stance. Nor does a balanced and realistic message fall within the growing mainstream or leading edge of popularity. If the message presented here is unpopular, then that is truly unfortunate—but if that is the case, the message is one that may be all the more important to hear. If you have found this book to have some measure of value, please suggest it to others who might also have the intellectual maturity to be open to balance and reason. The consequences of taking action without realistic clarity and without a balanced approach will only be to the detriment of our civilization.

NOTES

What are the most common greenhouse gases?

1. http://www.epa.gov/climatechange/emissions/downloads06/07introduction.pdf
2. Spahni R. et al., 2005, "Atmospheric Methane and Nitrous Oxide of the Late Pleistocene from Antarctic Ice Cores," *Science*, 310, 1317–1321, 25 November 2005. The pre-industrial concentrations are provided by ice core data.
3. Private communication with researchers for NOAA/ESRL atmospheric analysis done using weather balloons in Colorado and the surrounding region, 2007.
4. Stephens B. et al., 2007, "Weak Northern and Strong Tropical Land Carbon from Vertical Profiles of Atmospheric CO_2." *Science*, vol. 316, pp.1732–1735, 22 June 2007. doi: 10.1126/science.1137004
5. J. P. Peixoto and A. H. Oort, 1992, *Physics of Climate*, Figure 6.2, pp.92–93

Are all greenhouse gases created equal?

1. J. P. Peixoto and A. H. Oort, 1992, *Physics of Climate*, p.93.
2. http://brneurosci.org/c02.html
3. http://www.eia.doe.gov/cneaf/alternate/page/environment/appd_d.html
4. IPCC, 2001, Third Assessment Report
5. Spahni R. et al., 2005, "Atmospheric Methane and Nitrous Oxide of the Late Pleistocene from Antarctic Ice Cores," *Science*, 310, 1317–1321, 25 November 2005.
6. P. J. Michaels and R. C. Balling, 2000, *The Satanic Gases*, p.36

Is the Earth getting warmer? Is global warming real?

1. Al Gore, 2006, *An Inconvenient Truth*, August
2. IPCC, 2007, Fourth Assessment Report, FAQ, p.103
3. Santer B. et al., 2003, "Influence of Satellite Data Uncertainties on the Detection of Externally Forced Climate Change," *Science*, v01.300, 23 May 2003, pp.1280–1284, doi: 10.1126/science.1082393
4. Santer B. et al., 2005, "Amplification of Surface Temperature Trends and Variability in the Tropical Atmosphere," *Science*, v01.309, 2 September 2005, pp.1551–1556, doi: 10.1126/science.1114867
5. Ibid.
6. Thompson D., 2008, http://www.independent.co.uk/environment/climate-change/case-against-climate-change-discredited-by-study-835856.html
7. Walters J. T. et al., 2007, "Positive surface temperature feedback in the stable nocturnal boundary layer," *Geophysical Research Letters*, doi: 10.1029/2007GL029505
8. Christy J. R., 2007, Senate Commerce, Science and Transportation Committee, p.3, 14 November 2007, http://commerce.senate.gov/public/_files/christyjr_cst_071114_written.pdf

9. Santer B. et al., 2003, "Influence of Satellite Data Uncertainties on the Detection of Externally Forced Climate Change," *Science*, vol.300, 23 May 2003, pp.1280–1284, doi: 10.1126/science.1082393

10. Santer B. et al., 2005, "Amplification of Surface Temperature Trends and Variability in the Tropical Atmosphere," *Science*, vol.309, 2 September 2005, pp.1551–1556, doi: 10.1126/science.1114867

11. Spencer R. and Christy J. of University of Alabama at Huntsville

12. Mears C. et al., 2007, Remote Sensing Systems, Santa Rosa, CA

13. U.S. Climate Change Science Program

14. Mears C. and Wentz F., 2005, "The Effect of Diurnal Correction on Satellite-Derived Lower Tropospheric Temperature," *Science*, vol.309, pp.1548–1551, 2 September 2005

15. Santer B. et al., 2003, "Influence of Satellite Data Uncertainties on the Detection of Externally Forced Climate Change," *Science*, vol.300, 23 May 2003, pp.1280–1284, doi: 10.1126/science.1082393

16. Santer B. et al., 2005, "Amplification of Surface Temperature Trends and Variability in the Tropical Atmosphere," *Science*, vol.309, 2 September 2005, pp.1551–1556, doi: 10.1126/science.1114867

17. Christy J., 2007, "My Nobel Moment," *Wall Street Journal*, p.A19, November 1, 2007

18. IPCC, 2007, Fourth Assessment Report, FAQ, p.103

19. Santer B. et al., 2005, "Amplification of Surface Temperature Trends and Variability in the Tropical Atmosphere," *Science*, vol.309, 2 September 2005, pp.1551–1556, doi: 10.1126/science.1114867

20. http://www.aip.org/history/climate/gcm.htm#L_0295

21. IPCC, 2007, Fourth Assessment Report, Working Group 1 report, pp.597–599

22. Richard B. Alley, 2000, *The Two-Mile Time Machine*, Princeton University Press

23. Mark Bowen, 2005, *Thin Ice*, Henry Holt and Company

24. Scafetta N. and West B., 2006 (c), "Phenomenological solar signa-

ture in 400 years of reconstructed Northern Hemisphere temperature record," *Geophysical Research Letters*

25. Ibid.

26. The National Ice Core Laboratory, Denver, Colorado, a division of USGS, (located off of Kipling and 6th Avenue in Denver Federal Center, Building 53)

27. Mark Bowen, 2005, *Thin Ice*, Henry Holt and Company, p.394

28. Mark Bowen, 2005, *Thin Ice*, Henry Holt and Company, p.52

29. Ludwig K., 1992, "Mass Spectrometric ^{230}Th - ^{234}U - ^{238}U Dating of the Devils Hole Calcite Vein," *Science*, v01.258, pp.284–287, 9 October 1992

30. Landwehr J., 2002, "Ice core depth-age relation for Vostok D and Dome Fugi ^{18}O records based on the Devils Hole paleotemperature chronology," USGS Open-File Report 02–266, 53p

31. Winograd I., 2002, "The California Current Devils Hole, and Pleistocene Climate," *Science*, v01.296, p.7, 5 April 2002

32. Winograd I. et al., 1997, "Duration and Structure of the Past Four Interglaciations," *Quaternary Research*, v01.48, pp.141–154, September 1997

What factors impact climate, and how much global warming is due to the greenhouse gas CO_2?

1. Soon, Willie W. H., 2005, "Variable solar irradiance as a plausible agent for multidecadal variations in the Arctic-wide surface air temperature record of the past 130 years," *Journal of the American Geophysical Union*, v01.32, 27 August 2005, Harvard-Smithsonian Center for Astrophysics, Cambridge, Massachusetts

2. Ibid.

3. http://icecap.us/images/uploads/solar_changes_and_the_climate.pdf

4. P. J. Michaels and R. C. Balling, 2000, *The Satanic Gases*, p.36

5. Wilson R. and Hudson H., 1988, "Sun luminosity variations in solar cycle 21," *Nature*, v01.332, pp.810–812

6. Hoyt D. V. and Schatten K. H., 1993, "A discussion of plausible solar irradiance variations, 1700–1992," *Journal of Geophysical Research*, v. 98A, v01.18, pp.895–906

7. Hoyt D. and Schatten K., 1997, *The Role of the Sun in Climate Change*, Oxford University Press

8. Lean J., Beer J. and Bradley R., 1995, "Reconstruction of solar irradiance since 1610: implications for climate change," *Geophysical Research Letters*, v01.22, pp.3195–3198

9. Lean J., 2000, "Evolution of the Sun's spectral irradiance since the Maunder Minimum," *Geophysical Research Letters*, v01.27, pp.2425–2428

10. Fligge M. and Solanki S. K., 2000, "The solar spectral irradiance since 1700," *Geophysical Research Letters*, v01.27, pp.2157–2160

11. Nature.com, 2008, vol. 453, no. 7195, May 28, http://www.heatisonline.org/contentserver/objecthandlers/index.cfm?ID=6953&Method=Full

12. http://www.independent.co.uk/environment/climate-change/case-against-climate-change-discredited-by-study-835856.html

13. Lockwood M. and Stamper R., 1999, "Long-term drift of the coronal source magnetic flux and the total solar irradiance," *Geophysical Research Letters*, v01.26, pp.2461–2464

14. Scafetta N. and West B., 2006 (a), "Phenominological solar contribution to the 1900–2000 global surface warming," *Geophysical Research Letters*, v01.33, doi:10.1029

15. Lockwood M., Stamper R., and Wild M. N., 1999, "A doubling of the Sun's coronal magnetic field in the last 100 years," *Nature*, vol. 399, pp.436–439

16. Svensmark, Henrik, *The Proceedings of the Royal Society*, Series A, Oct. 2006

17. Shaviv N., 2005, "On climate response to changes in the cosmic ray flux and radiative budget," JGR-Space, v01.110, A08105, 23 August 2005

18. Lawrence Solomon, 2008, *The Deniers*, Richard Vigilante Books, p.145

19. Lawrence Solomon, 2008, *The Deniers*, Richard Vigilante Books, p.163

20. Lockwood M. and Fröhlich C., 2007, "Recent oppositely directed trends on solar climate forcings and the global mean surface air temperature," *The Proceedings of the Royal Society*, Series A

21. Scafetta N. and West B., 2007, "Phenominological reconstructions of the solar signature in the Northern Hemisphere surface temperature records since 1600," *Journal of Geophysical Research*, v01.112, 3 November 2007

22. NASA/JPL/Malin Space Science Systems

23. http://www.msss.com/mars_images/moc/c02_science_rel/

24. Wallace S. Broecker, 1995, *The Glacial World According to Wally*, Eldigio Press

25. Richard B. Alley, 2000, *The Two-Mile Time Machine*, Princeton University Press

26. Mark Bowen, 2005, *Thin Ice*, Henry Holt and Company

27. http://www.aip.org/history/climate/cycles.htm#L_0295

28. J. D. Hays, J. Imbrie, and N. Shackleton, 1976, "Variations in the Earths orbit: pacemaker of the ice ages," *Science*, vol. 194, pp.1121–1132, 10 December 1976

29. J. Imbrie and K. Palmer Imbrie, 1979, *Ice Ages: Solving the Mystery*, Enslow Publishers

30. Wallace S. Broecker, 1995, *The Glacial World According to Wally*, Eldigio Press

31. http://www.aip.org/history/climate/cycles.htm#L_0295

32. Richard B. Alley, 2000, *The Two-Mile Time Machine*, Princeton University Press, pp. 95–98

33. Mark Bowen, 2005, *Thin Ice*, Henry Holt and Company, pp.279–288

34. Lawrence Solomon, 2008, *The Deniers*, Richard Vigilante Books, pp.172–175

35. Wallace S. Broecker, 1992, *The Glacial World According to Wally*, draft copy, pp.54–61 and section IV-B

36. Fischer H. et al., 1999, "Ice Core Records of Atmospheric CO_2 Around the Last Three Glacial Terminations," *Science*, v01.283, n0.5408, pp.1712–1714, doi: 10.1126/science.283.5408.1712

37. Mudelsee M., 2001, "The phase relations among atmospheric CO_2 content, temperature, and global ice volume over the past 420 ka," *Quaternary Science Reviews*, 20:583–589

38. Caillon N. et al., 2003, "Timing of Atmospheric CO_2 and Antarctic Temperature Changes Across Termination III," *Science*, v01.299, pp.1728–1731, 14 March 2003

39. Richard B. Alley, 2000, *The Two-Mile Time Machine*, Princeton University Press

40. Mark Bowen, 2005, *Thin Ice*, Henry Holt and Company

41. Peter D. Ward, 2007, *Under a Green Sky*, Smithsonian Books

42. Christy J. R., 2007, Senate Commerce, Science and Transportation Committee, p.2, 14 November 2007, http://commerce.senate.gov/public/_files/christyjr_cst_071114_written.pdf

43. Spencer R., Braswell W., Christy J., Hnilo J., 2007, "Cloud and radiation budget changes associated with tropical intraseasonal Oscillations," *Geophysical Research Letters*, 9 August 2007, doi: 10.1029/200/GL029698

44. Svensmark, Henrik, 2006, *The Proceedings of the Royal Society*, Series A, October

45. Le Quéré C. et al., 2007, "Saturation of the Southern Ocean C02

Sink due to Recent Climate Change," *Science*, v01.316, pp.1735–1738, 22 June 2007, doi: 10.1126/science.1136188

46. IPCC, 2007, Fourth Assessment Report, working Group I report, Table 8.2, p.631

What impact does mankind have on increased atmospheric greenhouse gases?

1. http://www.epa.gov/climatechange/emissions/usinventoryreport.html
2. IPCC, 2007, Fourth Assessment Report, FAQ, p.105
3. Keppler F. et al., 2006, "Methane emissions from terrestrial plants under aerobic conditions," *Nature*, vol. 439, p.187–191, January 12, 2006, doi: 10.1038/nature04420
4. *New Scientist*, 26 March 2005, p. 20
5. Dueck T. et al., 2007, *New Phytologist*, April 27, doi: 10.1111/j.1469-8137.2007.02103.x

What about temperature projections, GHG thresholds, and expected outcomes?

1. Bjorn Lomborg, 2007, *Cool It*, Alfred A. Knopf Books, p.5
2. http://www.informs.org/article.php?id=1383
3. Peter D. Ward, 2007, *Under a Green Sky*, Smithsonian Books, p.179
4. Peter D. Ward, 2007, *Under a Green Sky*, Smithsonian Books, p.182
5. Mark Bowen, 2005, *Thin Ice*, Henry Holt and Company, p.393
6. Richard B. Alley, 2000, *The Two-Mile Time Machine*, Princeton University Press, p.92
7. Peter D. Ward, 2007, *Under a Green Sky*, Smithsonian Books, p.179

8. Bjorn Lomborg, 2007, *Cool It*, Alfred A. Knopf Books, p.63
9. http://www.ncpa.org/ba/ba282.html
10. Peter D. Ward, 2007, *Under a Green Sky*, Smithsonian Books, p.186
11. Peter D. Ward, 2007, *Under a Green Sky*, Smithsonian Books, p.179
12. Richard B. Alley, 2000, *The Two-Mile Time Machine*, Princeton University Press, p.92
13. http://pubs.usgs.gov/fs/fs2–00/
14. Mark Bowen, 2005, *Thin Ice*, Henry Holt and Company, p.252
15. Richard B. Alley, 2000, *The Two-Mile Time Machine*, Princeton University Press, p.92
16. Wallace S. Broecker, 1992, *The Glacial World According to Wally*, DRAFT COPY, section I, p.11
17. Bjorn Lomborg, 2007, *Cool It*, Alfred A. Knopf Books, p.70
18. IPCC, 2007, Fourth Assessment Report, FAQ, p.112 and p.114
19. Peter D. Ward, 2007, *Under a Green Sky*, Smithsonian Books, chapter 6
20. Peter D. Ward, 2007, *Under a Green Sky*, Smithsonian Books, pp.137–138
21. Michio Kaku, 1995, *Hyperspace*, Anchor Books, pp.296–297

Why do some believe that global warming may not necessarily be bad?

1. Robert Mendelsohn and James E. Neumann, 1999, *The Impact of Climate Change on the United States Economy*, Cambridge University Press

Can mankind take action to avoid a climate "jump" or avoid the next ice age?

1. Brian Fagan, 2004, *The Long Summer, How Climate Changed Civilization*
2. Peter D. Ward, 2007, *Under a Green Sky*, Smithsonian Books, pp.159–160

How much should human-caused greenhouse gas production be reduced?

1. Bjorn Lomborg, 2007, *Cool It*, Alfred A. Knopf Books, p.161
2. Bjorn Lomborg, 2007, *Cool It*, Alfred A. Knopf Books, p.23 and p.152
3. Bjorn Lomborg, 2007, *Cool It*, Alfred A. Knopf Books, p.22
4. Bjorn Lomborg, 2007, *Cool It*, Alfred A. Knopf Books, p.71
5. Bjorn Lomborg, 2007, *Cool It*, Alfred A. Knopf Books, pp.77–79
6. Bjorn Lomborg, 2007, *Cool It*, Alfred A. Knopf Books, p.100
7. Bjorn Lomborg, 2007, *Cool It*, Alfred A. Knopf Books, p.33
8. Ibid.
9. http://www.hm-treasury.gov.uk/media/4/3/Executive_Summary.pdf
10. Service, Robert F., 2004, "The Hydrogen Backlash," and "Choosing a CO_2 Separation Technology," *Science*, vol. 305, pp.962–963, 13 August 2004
11. José D. Figueroa, 2007, Carbon Capture and Sequestration Program Overview, DOE-NETL paper presented at the 2007 UOP Refining R&D Panel, May 2–4 in Santa Barbara, California
12. IPCC, 2007, Fourth Assessment Report, Working Group I report
13. IPCC, 2007, Fourth Assessment Report, Working Group II report
14. IPCC, 2007, Fourth Assessment Report, Working Group III report

15. IPCC, 2007, Fourth Assessment Report, Working Group I, Summary for Policymakers, p.12
16. IPCC, 2007, Fourth Assessment Report, Working Group III, Summary for Policymakers, p.14

What is the outlook for fossil fuels?

1. www.exxonmobil.com/corporate/files/energy_outlook_2007.pdf (November 2007)
2. George A. Olah, Alain Goeppert, and G. K. Surya Prakash, 2005, *Beyond Oil and Gas: The Methanol Economy*, Wiley-VCH, with permission

What can we reasonably do?

1. Bill T. Spence, 2007, Royal Dutch Shell Oil Company paper presented at the Shell Regional Symposium, November 7–8 in Panama, with permission
2. Cleaning up, a special report on business and climate change, 2007, *The Economist*, June 2
3. McKinsey and Company, 2007, private communication
4. www.eia.doe.gov/emeu/reps/enduse/er01_us_tab1.html
5. Christy J., 2007, "My Nobel Moment," *Wall Street Journal*, p.A19, November 1, 2007
6. Ibid.
7. Muse Stancil, 2008 analysis
8. Tom Ryan, 2007, Future Engine-Fuel Systems, Southwest Research Institute paper presented at the 2007 UOP Refining R&D Panel, May 2–4 in Santa Barbara, California, with permission
9. IPCC, 2007, Fourth Assessment Report, Working Group III report
10. Cleaning up, a special report on business and climate change, 2007, *The Economist*, June 2

11. www.exxonmobil.com/corporate/files/energy_outlook_2007.pdf, November 2007

12. Cleaning up, a special report on business and climate change, 2007, *The Economist*, June 2

13. George A. Olah, Alain Goeppert, and G. K. Surya Prakash, 2005, *Beyond Oil and Gas: The Methanol Economy*, Wiley-VCH, with permission

14. *The Hydrogen Economy: Opportunities, Costs, Barriers, and R&D Needs*, 2004, National Research Council and National Academy Engineering, The National Academy Press

15. Cleaning up, a special report on business and climate change, 2007, *The Economist*, June 2

16. IPCC, 2007, Fourth Assessment Report, Working Group III

17. R. W. Beck Corporation, 2007, private communication

18. DOE-EIA, 2007

19. The International Atomic Energy Agency, 2007

20. Cleaning up, a special report on business and climate change, 2007, *The Economist*, June 2

21. Service, Robert F., "The Hydrogen Backlash," and "Choosing a CO_2 Separation Technology," *Science*, 2004, vol. 305, pp.962–963, 13 August 2004

22. José D. Figueroa, 2007, Carbon Capture and Sequestration Program Overview, DOE-NETL paper presented at the 2007 UOP Refining R&D Panel, May 2–4 in Santa Barbara, California

23. www.exxonmobil.com/corporate/files/energy_outlook_2007.pdf, November 2007

24. The International Energy Agency, 2007

Should you believe everything you read?

1. Richard S. Lindzen, 2007 "Climate of Fear," *Wall Street Journal*, April 12

2. Willie Soon and Sallie Baliunas, 2003a, "Proxy climatic and environmental changes of the past 1000 years," *Climate Research*, v01.23, pp.89–110, January 31

3. Barbara W. Tuchman, 1984, *The March of Folly*, Alfred A. Knopf Inc.

4. Mann M. E., Bradley R. S., and Hughes M. K., 1998, "Global-scale temperature patterns and climate forcing over the past six centuries," *Nature*, v01.392, pp.779–787

5. *Energy & Environment*, 2003, v01.14, No. 6, p.766

6. McKitrick R., 2003, "What is the 'Hockey Stick' Debate About?," APEC Study Group, Department of Economics, University of Guelph, April 4

7. C. Essex and R. McKitrick, 2007, *Taken by Storm*, Key Porter Books, pp.175–183

8. Phillip Cooney, 2007, White House Filtered Climate Research, Scientists Says, *Oil Daily*, March 20, p.5

9. Chris Landsea, e-mail message to Kevin Trenberth and Linda Mearns, October 21, 2004, http//www.nuclear.com/archive/2005/01/20/20050120–001.html

10. Chris Landsea, e-mail message to Dr. R. Pachauri et al., November 5, 2004, http//www.nuclear.com/archive/2005/01/20/20050120–001.html

11. R. K. Pachauri, e-mail message to Chris Landsea, November 20, 2004, http//www.nuclear.com/archive/2005/01/20/20050120–001.html

Why is there so much emotion and division on the subject of global warming and climate change?

1. The Pew Research Center, 2007

2. http://people-press.org/reports/display.php3?reportid=303

3. http://www.americans-world.org/digest/global_issues/global_warming/gw1.cfm

4. Bjorn Lomborg, 2007, *Cool It*, Alfred A. Knopf Books, p.164

INDEX

STEVEN E. SONDERGARD